EMERGING MILITARY TECHNOLOGIES

MATTHEW N. O. SADIKU
SARHAN M. MUSA
PAUL A. ADEKUNTE

ISBN: 979-8-89419-025-9 (sc)
ISBN: 979-8-89419-026-6 (hc)
ISBN: 979-8-89419-027-3 (e)

THE EWINGS PUBLISHING

One Galleria Blvd., Suite 1900, Metairie, LA 70001
(504) 702-6708

DEDICATED TO OUR SIBLINGS:

Moses, David (late), James, and Julius
Rasmia (late), Majeda (late), Huda, and Osama
Emmanuel, Florence, Samson, and Bunmi

CONTENTS

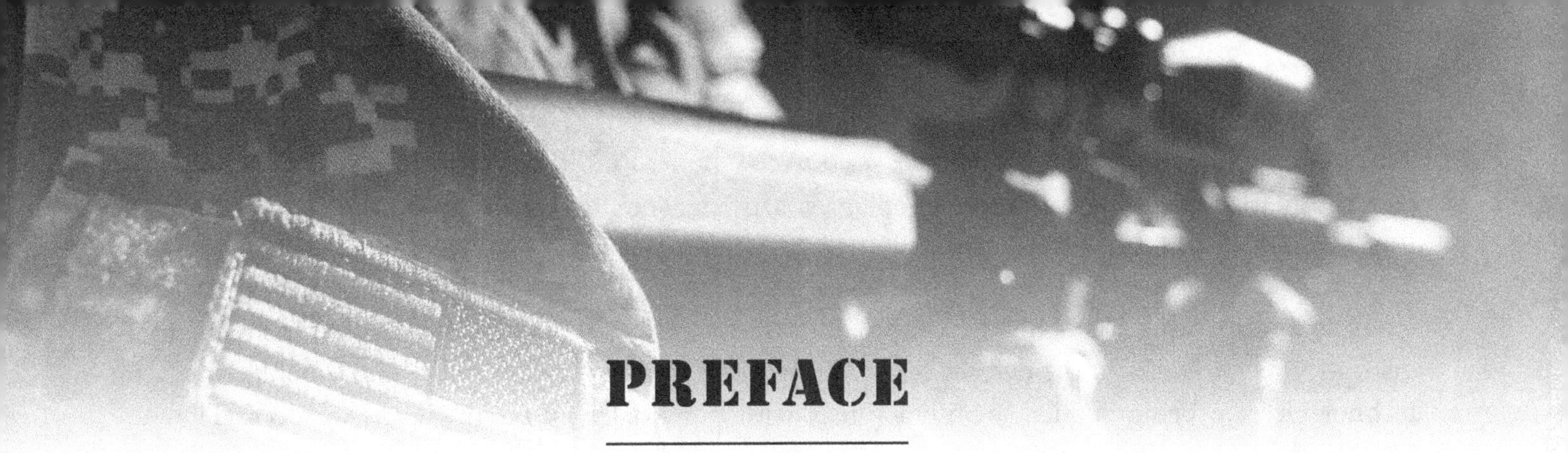

PREFACE

We live in the digital age where everything is touched and connected by technology. Our homes, our cars, and our jobs are all connected to technology. Modern societies are increasingly dependent on technology. Technology surrounds every aspect of 21st-century life. It is in the cell phones we use, the cars we drive, and even the food we eat. Technology has spanned the globe, connecting devices and people of all nations. It has become integrated into personal, professional, and social aspects of our lives. It is becoming more and more in demand in every sector of the economy, particularly in the military. The pace of change in military technology is unprecedented, but human nature does not change at these technological timescales.

The military, also known collectively as armed forces, is an armed and organized force primarily intended for warfare. The US military consists of seven branches that report directly to the Department of Defense (DoD). The military often looks to emerging technologies for new services or tools that will help them create a competitive advantage. The US military has long relied upon technological superiority to ensure its dominance in conflict and to underwrite US national security. The defense sector is at the forefront of technological innovation, driven by the need to maintain military superiority and safeguard national security. The global military apparatus is witnessing significant transformations and leveraging technology trends to strengthen capabilities.

An emerging technology may be regarded as an enhanced or completely new technology that brings about a radical change. It is a term that is often used to describe a new technology. Emerging technologies are shaping our societies. They are responsible for developing new products or devices. The emerging technologies discussed in this book have the potential to reshape the defense landscape and governments will need to stay abreast of the local developments if they are to maintain a strategic and operational advantage. Emerging military technology trends are changing the battlefield in four aspects—connectivity, lethality, autonomy, and sustainability. The US system created the world's most advanced military technology. The United States is the leader in developing many of the new technologies. However, China and Russia—key strategic competitors—are making steady progress in developing advanced military technologies.

This book explores emerging technologies used in the modern military. It is organized into 12 chapters that summarize emerging military technologies: artificial intelligence, robotics, drones, 3D printing, Internet of things, big data, Blockchain, cybersecurity, biotechnology, nanotechnology, and gamification

Chapter 1 - Introduction: This chapter provides an overview of some of the most prominent emerging disruptive technologies in the defense sector. It serves as an introduction to the entire book. Emerging technologies are being applied to military use, with potentially far-ranging consequences. They are capable of serving military missions for intelligence, surveillance, and reconnaissance. They will have a revolutionary impact on the battlefield of the future.

Chapter 2 - Artificial Intelligence in the Military: This chapter examines various applications of artificial intelligence (AI) in the military and defense. AI refers to computer systems that mimic human cognitive functions. AI has done remarkable things such as defeating human experts at various games. Militaries and defense organizations can use AI for autonomous weapons, autonomous vehicles, surveillance, cybersecurity, military intelligence, homeland security, logistics and transportation, military intelligence, and war planning. AI will have immense impact on national and international security.

Chapter 3 - Robotics: This chapter examines the various uses of robots in the military. Robotics is the discipline of designing and constructing intelligent machines, called robots. The integration of robotics in the military has ushered in a new era of efficiency, precision, and adaptability on and off the battlefield. Today's modern military forces are using different kinds of robots for different applications ranging from mine detection, surveillance, logistics, and rescue operations. The use of robots in warfare is being researched as a possible future means of fighting wars. Military robots save military lives by using these robots in applications that could be dangerous for human personnel.

Chapter 4 - Drones: In this chapter, we examine the applications of drones in the military. A drone may be regarded as a flying robot. Today, drones primarily serve the military industry around the world. Drones have revolutionized the way modern warfare is conducted by providing a means of engaging targets with precision and reduced risk to military personnel. Military drones have become indispensable tools for special operations forces, providing real-time situational awareness, communication relay, and electronic warfare capabilities. Drones are the future of warfare.

Chapter 5 - 3D Printing: This chapter takes a look at how the defense industry has embraced 3D printing. A 3D printer works by "printing" objects. 3D printing is being used by military units from different countries around the world. They are creating everything from replacement parts for critical vehicles, ships, and aircraft as well as new designs for safety equipment. 3D printing technology is revolutionizing the military industry by providing a way to quickly and cheaply produce spare parts, prototypes, and even entire weapons systems.

Chapter 6 - Internet of Things: This chapter introduces how the defense industry can leverage the opportunities created by the Internet of things (IoT). IoT is a worldwide network that connects devices to the Internet and to each other using wireless technology. It enables people and objects to interact with each other. The adoption of the IoT to military applications has a substantial impact on soldiers on the battlefield. Military IoT includes everything from battlefield sensors and weapons systems to tracking devices, communications equipment, wearables, drones, ships, planes, tanks, and even body sensors.

Chapter 7 - Big Data: In this chapter, we will delve into the role of big data in military operations and explore the realm of possibilities in military data analytics. Big data refers to data sets of extreme size which are beyond the capability of the commonly used software tools. It is regarded as the most strategic resource of the 21st century, similar in importance to gold and oil. It is used in a variety of battlefield functions, such as targeted killing operations and intelligence collection and analysis. It is now a key tool to investigate and prosecute those responsible for wartime atrocities.

Chapter 8 - Blockchain: In this chapter, we explore how Blockchain could play a pivotal role in shaping the future of the military army. Blockchain is a peer-to-peer network that sits on top of the Internet. Integration of Blockchain into military operations can significantly improve security, resource allocation, fraud reduction, and operational resilience. Blockchain technologies can also support food safety and health care challenges on the battlefield, build health data sharing platforms for increased security and efficiency, track, and trace the food supply chain to prevent food related outbreaks.

Chapter 9 - Cybersecurity: This chapter introduces readers to cybersecurity in the military domain. Cybersecurity refers to a set of behaviors, methods, and technologies aimed at protecting systems, networks, data, and computers from harm, attacks, and illegal access. Military cybersecurity refers to the practice of protecting military organizations, assets, and operations from cyber threats and attacks. It is critical in protecting both their troops and the citizens. Its operations are shifting to a digital battlefield, where tools and technology work to save lives and increase efficiency.

Chapter 10 - Biotechnology: This chapter examines the various uses of biotechnology in the military. Biotechnology is technology that utilizes biological systems or living organisms to develop products. It pervades almost all aspects of our daily life; it affects the foods we eat, the safety of the water we drink, the clothes we wear, the medications we take, etc. Biotechnology and the military are strengthening the power of soldiers and resisting fatigue. It senses and monitors the battlefield. Although biotechnology is one of the most versatile, exciting, and innovative technologies of the 21st century, its benefits for defense have yet to be fully explored.

Chapter 11 - Nanotechnology: This chapter focuses on the use of nanotechnology in various military applications. Nanotechnology is the study and manipulation of matter at incredibly small sizes. It has a great role to play in materials and systems for military use. The use of nanotechnology in the area of warfare and defense has been rapid and expansive. The main goals of military research into nanotechnology are to improve medical and casualty care for soldiers, and to produce lightweight, strong and multi-functional materials for use in clothing.

Chapter 12- Gamification: In this chapter, we provide an overview on gamification in the military. Game has become a significant part of human's culture. Gamification involves taking some of the things that make gaming attractive, addictive, interesting and challenging, and applying them to those non-gaming areas such as business, education, the military, etc. Gamification in the military training programs is becoming more engaging, effective, and efficient. Gamification and 3D training significantly increase combat effectiveness by increasing the level of training provided to military personnel who operate the equipment.

This book is a comprehensive text on the emerging technologies in the military. It provides an overview of each emerging technology in a way that beginners can understand. It is a must read for those interested in the military and its future.

—M. N. O. Sadiku, S. M. Musa, and P. A. Adekunte

ABOUT THE AUTHORS

A. Matthew N. O. Sadiku received his B. Sc. degree in 1978 from Ahmadu Bello University, Zaria, Nigeria and his M.Sc. and Ph.D. degrees from Tennessee Technological University, Cookeville, TN in 1982 and 1984 respectively. From 1984 to 1988, he was an assistant professor at Florida Atlantic University, Boca Raton, FL, where he did graduate work in computer science. In total, he received seven college degrees. From 1988 to 2000, he was at Temple University, Philadelphia, PA, where he became a full professor. From 2000 to 2002, he was with Lucent/Avaya, Holmdel, NJ as a system engineer and with Boeing Satellite Systems, Los Angeles, CA as a senior scientist. He is presently a Regents professor emeritus of electrical and computer engineering at Prairie View A&M University, Prairie View, TX.

He is the author of over 1,230 professional papers and over 130 books including "Elements of Electromagnetics" (Oxford University Press, 7th ed., 2018), "Fundamentals of Electric Circuits" (McGraw-Hill, 7th ed., 2020, with C. Alexander), "Computational Electromagnetics with MATLAB" (CRC Press, 4th ed., 2019), "Principles of Modern Communication Systems" (Cambridge University Press, 2017, with S. O. Agbo), and "Emerging Internet-based Technologies" (CRC Press, 2019). In addition to the engineering books, he has written Christian books including "Secrets of Successful Marriages" (with J. O. Sadiku), "How to Discover God's Will for Your Life," and commentaries on all the books of the New Testament Bible. Some of his books have been translated into French, Korean, Chinese (and Chinese Long Form in Taiwan), Italian, Portuguese, Spanish, German, Dutch, Polish, and Russian.

He was the recipient of the 2000 McGraw-Hill/Jacob Millman Award for outstanding contributions in the field of electrical engineering. He was also the recipient of Regents Professor award for 2012-2013 by the Texas A&M University System. He is a registered professional engineer and a life fellow of the Institute of Electrical and Electronics Engineers (IEEE) "for contributions to computational electromagnetics and engineering education." He was the IEEE Region 2 Student Activities Committee Chairman. He was an associate editor for IEEE Transactions on Education. He is also a member of Association for Computing Machinery (ACM). His current research interests are in the areas of computational electromagnetic, computer science/networks, engineering education, and marriage counseling. His works can be found in his autobiography, "My Life and Work" (Trafford Publishing, 2024) or his website: www.matthew-sadiku.com. He can be reached via email at sadiku@ieee.org.

B. Sarhan M. Musa is the Associate Dean of Graduate Studies at Prairie View A&M University. He earned his PhD in Electrical Engineering from the City University of New York. Additionally, he is the founder and director of the Prairie View Networking Academy (PVNA) in Texas. Professor Musa is recognized as an LTD Sprint and Boeing Welliver Fellow. His research and scholarly contributions have garnered international recognition, and he has delivered numerous invited presentations at global conferences. He has been honored with several prestigious national and university awards, as well as research grants. A senior member of the IEEE, he has also participated as a member of the technical program and steering committees for various prominent journals and conferences. Professor Musa has authored over a dozen books covering diverse topics within electrical and computer engineering. His current research interests encompass various subjects, including artificial intelligence and machine learning, renewable energy, power systems, control systems, and computational methods.

C: Paul A. Adekunte received his B.Sc in zoology from Ahmadu Bello University, Zaria, Nigeria in 1983 and masters degree in crime management and prevention from Bayero University, Kano, Nigeria in 2011. He got his postgraduate degree in Biology in 2005 from the Federal College of Education (Technical) Akoka, Lagos, Nigeria. He also obtained certificates in public relations and computer studies. He is a professional educator in biology at the secondary and advanced levels as well a security expert. He is a member of several professional bodies. His current areas of research interests include management, educational guidance and counseling, conflict and crisis resolution, and security matters. He has published several professional papers. He currently resides with his wife in Lagos, Nigeria and can be reached via email at adekuntepaul@gmail.com.

CHAPTER 1

INTRODUCTION

"The art of war is simple enough. Find out where your enemy is. Send drones out as soon as you can. Dither about sending in ground forces as hard as you can, and keep passing the buck to someone else."
– Ulysses S. Grant

1.1 INTRODUCTION

We live in the digital age where everything is touched and connected by technology. Our homes, our cars, and our jobs are all connected to technology. Modern societies are increasingly dependent on technology. Technology is everywhere. It surrounds every aspect of 21st-century life. It is in the cell phones we use, the cars we drive, and even the food we eat. Technology has spanned the globe, connecting devices and people of all nations. It has become integrated into personal, professional, and social aspects of our lives. It is getting better, smaller, and faster. It is becoming more and more in demand in every sector of the economy, particularly in the military. The pace of change in military technology is unprecedented, but human nature does not change at these technological timescales.

The US military has long relied upon technological superiority to ensure its dominance in conflict and to underwrite US national security. The defense sector is at the forefront of technological innovation, driven by the need to maintain military superiority and safeguard national security. The global military apparatus is witnessing significant transformations and leveraging technology trends to strengthen capabilities. Emerging military technology trends are changing the battlefield in four aspects—connectivity, lethality, autonomy, and sustainability. Members of Congress and Pentagon officials are increasingly focused on developing emerging military technologies to enhance US national security and keep pace with US competitors [1].

This chapter provides an overview of some of the most prominent emerging disruptive technologies in the defense sector. It begins with discussing the military. It explains what we mean by emerging technologies.

It covers twelve popular emerging military technologies. It highlights the benefits and challenges of these technologies. It concludes with comments.

1.2 THE MILITARY

The defense sector is constantly evolving, driven by technological advancements and the need to adapt to emerging threats. Crucial objectives for militaries include protecting forces, increasing situational awareness, reducing soldiers' workload, and facilitating movement in challenging terrains.

Figure 1.1 US Forces [2].

The United States military has several components or branches, as shown in Figure 1.1 [2]. The seven branches of the US military are the Army, Navy, Air Force, Marines, Coast Guard, National Guard, and Space Force. The Army is the oldest branch, and the Space Force is the newest. The Army National Guard and the Air National Guard are reserve components of the Army and Air Force, respectively. For example, the US Coast Guard is depicted in Figure 1.2 [3].

Figure 1.2 US Coast Guard [3].

The branches all report to the Department of Defense (DoD). Regardless of branch, the Army conducts both operational and institutional missions. The operational Army consists of numbered armies, corps, divisions, brigades, and battalions that conduct full-spectrum operations around the world. Institutional organizations provide the infrastructure necessary to raise, train, equip, deploy, and ensure the readiness of all Army forces [4]. Some military personnel work in the air, some on the ground or water, and others in space.

The US system created the world's most advanced military. The United States is the leader in developing many of the new technologies. However, China and Russia—key strategic competitors—are making steady progress in developing advanced military technologies. As these technologies are integrated into foreign and domestic military forces and deployed, they could hold significant implications for the future of international security.

The US military is racing to keep up with advances by China and Russia in hypersonic weapons, which travel at five times the speed of sound or faster. In contrast to Russia and China, the United States is not known to be developing hypersonic weapons for potential use with a nuclear warhead. As a result, the United States is seeking to develop hypersonic weapons that can attack targets with greater accuracy. China's defense transformation has been guided by a principle known as "military-civil fusion," which aims to allow the state to seamlessly capitalize on private-sector advances. The United States has a finite window to up its game against China, which already uses AI in a vast domestic surveillance network and has staked out a goal of AI primacy by 2030. A factor slowing down US innovation is the Pentagon's focus on long-term investments in a small number of weapons systems, some of which do not play out as planned. China, meanwhile, tends to experiment with many versions of similar technology. Figure 1.3 portrays Chinese army [5].

Figure 1.3 Chinese army [5].

As we approach 2040, the arena of warfare and defense stands on the brink of a radical shift, driven by swift advancements in military technology. NATO expects four core characteristics to define many key advanced military technologies [6]:

- *Intelligent* – Solutions will exploit integrated AI, knowledge-focused analytic capabilities and symbiotic AI-human intelligence, resulting in the spread of disruptive applications across the technological spectrum.

- *Interconnected* – Solutions will take advantage of the growing network of virtual and physical domains. Connectivity solutions address concerns about how combatants detect and locate their adversaries, communicate with each other, and direct operations. This will include networks of sensors, organizations, individuals and autonomous agents, linked via new encryption methods and distributed ledger technologies.

- *Distributed:* Solutions will employ decentralised and ubiquitous large-scale sensing, storage, and computation to achieve new disruptive military effects.

- *Digital:* Solutions will blend human, physical and information domains to support novel disruptive effects.

The introduction of the technologies on future battlefields will transform every aspect of combat and raise a host of challenges for advocates of responsible arms control.

1.3 WHAT ARE EMERGING TECHNOLOGIES?

Technology may be regarded as a collection of systems designed to perform some function. It can help alleviate some of the challenges facing business today. Emerging technology is a term generally used to describe new technology. The term often refers to technologies currently developing or expected to be available within the next five to ten years. Any imminent, but not fully realized, technological innovations will have some impact on the status quo.

An emerging technology may be regarded as an enhanced or completely new technology that brings about a radical change. Emerging technologies are shaping our societies. They continue to affect the way we live, work, and interact with one another. Emerging technology (ET) lacks a consensus on what classifies them as "emergent." It is a relative term because one may see a technology as emerging and others may not see it the same way. It is a term that is often used to describe a new technology. A technology is still emerging if it is not yet a "must-have" [7]. An emerging technology is the one that holds the promise of creating a new economic engine and is trans-industrial. ET is used in different areas such as media, healthcare, business, science, education, or defense.

The characteristics of emerging technologies include the following [8]:

- *Novelty:* Emerging technologies are typically new or novel, meaning they have yet to be widely adopted or used. They often represent a significant departure from existing technologies or processes.

- *Potential for Disruption:* Emerging technologies have the potential to disrupt existing markets, industries, or ways of doing things. They may also displace existing businesses or industries.

- *Uncertainty:* Because emerging technologies are still in the early stages of development, there is often a high uncertainty surrounding their future potential and impact. It can be challenging to predict how they will evolve.

- *Rapid Change:* Emerging technologies often evolve rapidly, with new developments and innovations emerging frequently. It can make keeping up with the latest trends and advancements challenging.

- *Interdisciplinary*: Emerging technologies often involve multiple disciplines or fields of study, such as computer science, engineering, and biology. They may require collaboration across different fields and industries to develop their potential fully.

Emerging technologies are worth investigating. They are responsible for developing new products or devices. The military often looks to emerging technologies for new services or tools that will help them create a competitive business advantage.

1.4 EMERGING MILITARY TECHNOLOGIES

Emerging military technologies are rapidly changing defence sector. This section presents some selected emerging military technologies in the United States, China, and Russia: artificial intelligence, robotics and autonomous weapons, Internet of things, cybersecurity, additive manufacturing, big data analytics, quantum computing, biotechnology, immersive technologies, drones, and 5G connectivity. These technologies are being applied to military use, with potentially far-ranging consequences. The consequences of failure on any of these technologies are tremendous—they could make the difference between victory and defeat. These technologies explained as follows [7,9]:

1. *Artificial Intelligence* (AI): AI is rapidly transforming the defense sector, offering enhanced capabilities when it comes to autonomous systems, data analysis, and decision-making. In some contexts, militaries will use AI to reduce the scope for human error in military exercises by automating processes. AI will play a key role in defense sector infrastructure. It will leverage predictive capabilities to manage maintenance for military assets, direct technology development, and assist with product testing. AI is currently being incorporated into a number of military

applications by both the United States and its competitors. In military and defense, AI adoption enhances computational reasoning for intelligence, surveillance, and reconnaissance missions. This advancement empowers autonomous weapon systems and ensures equipment safety, thereby reducing soldier casualties. Figure 1.4 shows autonomous weapons and artificial intelligence in warfare [10].

Figure 1.4 Autonomous weapons and artificial intelligence in warfare [10].

2. *Robotics and Autonomous Weapons Systems:* These are possibly the most controversial technological defense sector development for quite some time. The debate over autonomous weapons systems and the extent to which they should be independent of human intervention is still a major sticking point. The systems allow for the removal of human agency from decision-making processes, but they also possibly open the door to a new, less transparent, and less accountable form of warfare. Robotics and autonomous weapon systems improve the combat effectiveness of the military as well as impact other trends in the industry. Drones, a part of this system, enhance battlefield situational awareness. Multi-mission autonomous military vehicles are instrumental in landmine clearance, search and rescue operations, explosive ordnance disposal, and logistics support. Robots also have capabilities that humans do not have: the ability to stay awake 24/7, the ability to see from all angles, the ability to process information in an instant, etc. Robotic dogs are being used alongside Air Force service members. Figure 1.5 shows an autonomous weapons system [11].

Figure 1.5 An autonomous weapons system [11].

3. *Internet of Things* (IoT): The emergence of the IoT as a network of interconnected devices that can exchange data and perform tasks without human intervention will generate vast amounts of valuable data. This will provide militaries with greater insight into operational environments, capabilities, and performance. Technologies like wearable devices, sensors, and drones can all feedback insights, resulting in a more complex and comprehensive account of what is happening on the ground in any given location. There is a substantial rise in the use of the IoT technologies such as sensors, wearables, and edge computing. IoT applications in defense connect ships, planes, tanks, drones, soldiers, and operating bases into a cohesive network. This connection enhances perception, field understanding, situational awareness, and response time. Sensing and computing devices, worn by soldiers and embedded in their equipment, collect a variety of biometric data in the Internet of military things (IoMT) framework.

4. *Cybersecurity:* This is the protection of cyberspace from malicious attack. It has become a major national security priority. Cybersecurity, as perceived by US leaders, can take two forms: (1) defensive action aimed at protecting one's own information infrastructure against attack; and (2) offensive action intended to punish, or retaliate against, an attacker by severely disrupting its systems, or to deter such attack by holding out the prospect of such punishment. Secure operations in cyberspace has become essential for the continued functioning of the international economy. An extraordinary tool for many purposes, the Internet is also vulnerable to attack by hostile intruders, whether to spread misinformation, disrupt vital infrastructure, or steal valuable data. As digital connectivity increases, the defence sector faces an ever-growing threat from cyberattacks. Disruptive technologies in cybersecurity include advanced encryption methods, blockchain

technology for secure data sharing, and artificial intelligence-based threat detection systems. Vulnerability to cyber-attacks is a significant concern for military systems, as it could lead to the loss of classified information and damage to systems. Prescriptive security technology, utilizing cybersecurity, AI, and automation, detects potential threats and neutralizes them before they impact defensive cyber warfare capabilities. Advanced firewalls and intrusion detection systems play a crucial role in protecting sensitive military data and ensuring the integrity and security of military operations. As shown in Figure 1.6, cyber attacks are borderless [12].

Figure 1.6 Cyber attacks are borderless [12].

5. *3D Printing:* Improving performance in speed, capacity, and fuel consumption is crucial, and reducing the weight of defense equipment plays a significant role in this. 3D printing, also known as additive manufacturing, enables the production of components and parts using less material than traditional methods. It employs computer-aided design and computer-aided manufacturing capabilities to create objects through deposition, or layer-by-layer printing. It could transform the way we think about defense-sector supply chains and logistics. It allows for rapid prototyping, on-demand production of spare parts, and equipment customization. The technology reduces the time and cost required for manufacturing, increases flexibility and enables the production of complex geometries that are difficult to achieve through traditional manufacturing processes. It enables localized, on-demand production, thereby reducing the logistical burden. The US Navy is utilizing additive printing to ease supply chain issues. Figure 1.7 displays how 3D printing is used by the military [13].

Figure 1.7 3D printing is used by the military [13].

6. *Quantum Computing:* This is arguably the technology that has had the least impact on the defense sector. Quantum technology has not yet reached maturity. It has the potential to revolutionize cryptography and computational power. Quantum technology leverages unique quantum physics properties for potentially major improvements in cryptography, computing, sensing and communication security – all crucial for military operations. Quantum computers can solve complex problems exponentially faster than traditional computers and are extremely sensitive to environmental changes, ensuring they offer a significant advantage in target identification, weather forecasting, and data analysis. Quantum technology could have other military applications, such as quantum sensing, which could theoretically enable significant improvements in submarine detection, rendering the oceans "transparent." Quantum computing may allow adversaries to decrypt information, which could enable them to target US personnel and military operations. Quantum computing, applied in cryptanalysis and simulations, aids informed decision-making. Quantum technology may be far from making sizable impacts within military application, but its implications could be tremendous in terms of sensing, encryption, and communication.

7. *Biotechnology:* This leverages life sciences for technological applications. A number of developments in biotechnology hold potential implications for national security. In the US, unclassified use of biotechnology (i.e. using life sciences to support technological applications) has been confined to improving military readiness, resiliency, and recovery. The US has also sought applications that would improve solider strength, cognition, and endurance that would give them leverage over other humans. Only the weaponization of biotechnology is barred by international law.

8. *Immersive Technologies:* Building replicable and flexible experiences becomes easier with immersive technologies, especially for flight or combat training. Augmented reality (AR) goes beyond virtual reality (VR) in military training opportunities. It equips on-field soldiers with wearable glasses or AR headsets, providing mapping information, movement markers, and other data. This

technology enhances real-time decision-making for ground forces. In addition to these, immersive technologies assist in mission planning.

9. *Big Data:* In the evolving landscape of warfare, the role of data and its analysis is becoming increasingly crucial. Harnessing big data analytics, militaries unlock insights from diverse data sources, gaining a strategic edge. Efficient interpretation of data from the Internet of military things (IoMT) is another benefit of analytics.

10. *5G Connectivity:* In military operations, the importance of timely and appropriate information is paramount. Accelerating real-time decision support, 5G offers hyper-converged connectivity and secure data networks. 5G networks enable the transfer of vast data amounts to remote sensors and weapons, creating dense, resilient battlefield networks. It also ensures efficient remote control by offering seamless connectivity for unmanned systems, including drones and autonomous vehicles.

11. *Blockchain:* Data security is a crucial aspect of data sharing, and blockchain technology addresses this need effectively. Defense companies are developing solutions based on blockchain to safeguard confidential military data and combat cyber threats. This technology finds use in device tracking, procurement process streamlining, and supply chain security. Blockchain technology also ensures the integrity of internal elections and surveys by facilitating secure voting systems for military personnel.

12. *Drones*: The main reason behind the increasing use of drones in the military may be economics. Drones can be used to suppress missile and air attacks. They are exceedingly cheap to produce when compared to most other weapons. Cheap drones can easily destroy many expensive, high-tech weapons on the battlefield. With more drones being produced specifically for warfare, we are seeing new counter-drone technologies spring up. Figure 1.8 shows some military drones [14].

Figure 1.8 Some military drones [14].

1.5 BENEFITS

Emerging technologies are capable of serving military missions for intelligence, surveillance, and reconnaissance. In a time of escalating global tension, the technological innovations provide an essential lever for both defense and deterrence. Advances in missile and weapons technologies increase lethality, making battlefield operations more effective. The introduction of emerging technologies on future battlefields will transform every aspect of combat and raise a host of challenges for advocates of responsible arms control. The US Department of Defense (DOD) addresses means of rapidly identifying, adopting, prototyping, and fielding dual-use commercial technology. It is committed to strengthening its technical advantage in a strategic environment that is increasingly being shaped by high-tech competition. Other benefits of emerging military technologies include:

1. *Warfare Capabilities:* Imagine the Army having unmanned air and ground distribution platforms, the capability to manufacture replacement parts on the battlefield, and the ability to produce water from air. It imperative that the US not only seize the opportunity that artificial intelligence presents for warfighters, but to lead in its responsible development. The US has also introduced a political declaration on the responsible military use of artificial intelligence. Directed energy weapons have become much smaller and lighter.

2. *Increased Spending:* The military's emphasis on artificial intelligence and robotics is growing. Investments in advanced technologies such as AI, in addition to the department's ability to leverage capabilities of US allies, is critical for deterring and preventing conflict in the future.

3. *Advanced Defense:* The innovative technologies aim to provide better protection while reducing the weight burden on soldiers, thus improving mobility and comfort during extended operations. Many analysts believe that AI will revolutionize warfare by allowing military commanders to bolster or replace their personnel with a wide variety of "smart" machines.

4. *Efficient Operations:* AI can make warfare systems more efficient, which can reduce the need for human input and maintenance. This can also reduce the impact of human error. New technologies, such as augmented reality and heads-up displays, can help improve situational awareness and connectivity.

5. *Increased Efficiency:* Automation and AI can streamline logistics, maintenance, and operational planning, leading to more efficient use of resources.

6. *Training:* Virtual reality and simulation technologies provide realistic training environments, improving readiness without the costs of live exercises.

7. *Situational Awareness*: Technologies such as drones, satellites, and reconnaissance systems provide commanders with comprehensive situational awareness, helping in decision-making.

8. *Automation*: Developments in technologies such as AI, big data analytics, and lethal autonomous weapons could diminish or remove the need for a human operator. This could, in turn, increase combat efficiency and accelerate the pace of combat—potentially with destabilizing consequences.

9. *Decision-Making*: AI technologies can enable predictive analytics and autonomous systems, which can help improve decision-making.

1.6 CHALLENGES

The challenges of emerging technologies in the military include potential for unintended escalation due to rapid decision-making by AI systems, difficulty in managing complex technological integration, ethical concerns around autonomous weapons, lack of human oversight in critical situations, cybersecurity vulnerabilities, the rapid pace of technological evolution making it hard to keep up, and concerns about proliferation of these technologies to adversaries. There is a lack of agreement in the Congress on the priorities among emerging military technologies. This has led some analysts to suggest that DOD should adopt a technology strategy to set spending priorities that can be sustained over time. The commercial companies that are often at the forefront of innovation in emerging technologies are reluctant to partner with DOD due to the complexity and secrecy of the defense acquisition process. Other challenges of emerging military technologies include:

1. *Cost-effectiveness:* Some analysts have raised concerns about the potential operational risks posed by lethal autonomous weapons and directed energy weapons. What mission(s) will these weapons be used for? Are the weapons the most cost-effective means of executing these potential missions? In what circumstances and for what purposes should the US military's use of the weapons be permissible? For example, an autonomous weapon could continue engaging inappropriate targets and this could result in mass civilian casualties.

2. *Bias:* AI can often have unintended effects like racial bias due to algorithms developed with an noncomprehensive data set. Researchers have repeatedly discovered instances of racial bias in AI facial recognition programs due to the lack of diversity in the images on which the systems were trained, while some natural language processing programs have developed gender bias. Such biases could hold significant implications for AI applications in a military context.

3. *Ethical Concerns:* The development and use of autonomous weapons systems raise ethical questions about who is responsible for actions taken by these systems, especially when they involve lethal force. Is the US military appropriately balancing the potential warfighting utility of biotechnologies with ethical considerations? What, if any, national and international frameworks are needed to consider the ethical, moral, and legal implications of military applications of biotechnologies such as synthetic biology, genome editing, and human performance modification? For example, the use of fully autonomous weapons in combat automatically raises questions about the military's ability to comply with the laws of war and international humanitarian law, which require belligerents to

distinguish between enemy combatants and civilian bystanders. Even more worrisome, some of the weapons now in development, such as unmanned anti-submarine wolfpacks, could theoretically endanger the current equilibrium in nuclear relations among the major powers.

4. *Salary Disparity:* Some reports indicate that DOD and the defense industry have difficulty recruiting and retaining personnel with expertise in emerging technologies because research funding and salaries significantly lag behind those of commercial companies. Some personnel work in the air, some on the ground or water, and others in space.

5. *Safety*: Although robotics technology offers numerous opportunities, but there are concerns. The most obvious is that military robots can assume risks that could potentially lead to human casualties. We should underscore the degree to which progress in computers could create vulnerabilities, as nations increasingly utilized computer systems and software that created potentially gaping weaknesses in their military capabilities.

6. *Secrecy:* The US military uses communication jammers as part of electronic warfare, but many of the programs remain shrouded in secrecy. Official details are sparse.

7. *Competition:* Motivated by increasingly stiffer competition, the US is investing heavily in military weaponry that pushes the boundaries of technology. The US has remained a leader in those endeavors, with China and Russia closely following behind. If the major powers are prepared to discuss binding restrictions on the military use of destabilizing technologies, certain priorities take precedence. US is increasingly focused on developing emerging military technologies to enhance its national security and keep pace with its competitors. Although China has not been involved in real, hardcore combat for a few decades, the regular showcasing of its military might have created this aura of China being the undisputed military leader in key technological domains.

8. *Threat:* The deployment of fully autonomous weapons systems poses numerous challenges to international security and arms control, beginning with a potentially insuperable threat to the laws of war and international humanitarian law. Autonomous weapons systems could pose a potential threat to nuclear stability by investing their owners with a capacity to detect, track, and destroy enemy submarines and mobile missile launchers. Even the mere existence of such weapons could jeopardize stability. Today's stability rests on the belief that each major power possesses at least some devastating second-strike.

9. *Cyberattacks*: Integrating new technologies can introduce new vulnerabilities to cyberattacks, potentially disrupting critical military operations. Hackers and cyber terrorists, acting in official capacities for other countries or acting on their own, have shown the ability to take down critical infrastructures like electric grids and communications systems. The cyber domain is coming to resemble the strategic nuclear realm, with notions of defense, deterrence, and assured retaliation initially devised for nuclear scenarios now being applied to conflict in cyberspace. The stakes are too high.

10. *Cyberwar*: Today, a whole new array of technologies is being applied to military use, with potentially far-ranging consequences. Although the risks and ramifications of these weapons are not yet widely recognized, policymakers will be compelled to address the dangers posed by innovative weapons technologies and to devise international arrangements to regulate or curb their use. AI-powered weapons could make decisions too quickly, leading to unintended escalation of conflicts beyond human control, potentially triggering catastrophic events like nuclear war. Fighting cyberwar is not typical warfare. There is an abundance of ambiguity and deniability, which hinder defense officials from preventing attacks and making quick decisions once an attack has been launched. Some predict that cyber attackers will have weaponized operational technology in an effort to harm or kill humans.

Addressing these challenges will not be easy, but both current and future generations must contrive novel solutions to new perils. It will require interdisciplinary research teams that have the capacity to work across the physical, digital, and biological boundaries while collaborating seamlessly with end-users, human combatants.

1.7 CONCLUSION

DOD's ability to accelerate innovation and adoption of key technologies is a critical piece of the United States' strategy to build an enduring military advantage. The technology innovation strategies of the cold war era focused on R&D of large-scale hardware and weapons systems. As today's threat environment shifts, the United States needs a more strategic approach to successfully harness the technologies of the future.

The emerging technologies discussed in this chapter have the potential to reshape the defense landscape and governments will need to stay abreast of the local developments if they are to maintain a strategic and operational advantage. As these technologies are integrated into foreign and domestic military forces and deployed, they could hold significant implications for the future of international security. Rapid advances and convergence in fields such as artificial intelligence and robotics will continue to have a revolutionary impact on the battlefield of the future. More information on emerging military technologies is available from the books in [15-26] and the following related magazines:

- *Military Review*
- *Technology*

REFERENCES

[1] M. N. O. Sadiku, P. A. Adekunte, and J. O. Sadiku, "Emergent Military Technologies," *International Journal of Trend in Scientific Research and Development*, vol. 8, no. 5, September-October 2024, pp. 145-152.

[2] "Our Forces,"
https://www.defense.gov/About/our-forces/#:~:text=The%20Army%2C%20Marine%20Corps% 2C%20 Navy,in%20part%20under%20state%20authority

[3] "United States Coast Guard,"
https://www.history.uscg.mil/Our-Collections/Photos/igphoto/2003509414/

[4] "Understanding the Army's structure,"
https://home.army.mil/novosel/application/files/9016/2681/2207/Enclosure 3 FY21 Army Command Structure.pdf

[5] "Over 2,000 Chinese troops join Russian military drills,"
https://www.aa.com.tr/en/politics/over-2-000-chinese-troops-join-russian-military-drills/2674357

[6] "Exploring emerging disruptive technologies in the defence sector," July 2023,
https://www.flysight.it/exploring-emerging-disruptive-technologies-in-the-defence-sector/

[7] M. Halaweh, "Emerging technology: What is it?" *Journal of Technology Management & Innovation*, vol. 8, no. 3, 2013, pp. 108-115.

[8] N. Duggal, "Top 18 new technology trends for 2023," July 2023,
https://www.simplilearn.com/top-technology-trends-and-jobs-article

[9] 'Top 10 military technology trends & innovations for 2025,"
https://www.startus-insights.com/innovators-guide/top-10-military-technology-trends-2022/

[10] "Future weapons: AI, drones, and hypersonics by 2040,"
https://insiderrelease.com/future-of-weapons/

[11] "'Killer robots': the danger of lethal autonomous weapons systems,"
https://southernvoice.org/killer-robots-the-danger-of-lethal-autonomous-weapons-systems/

[12] M. T. Klare, "The challenges of emerging technologies," December 2018,
https://www.armscontrol.org/act/2018-12/features/challenges-emerging-technologies

[13] C. Schwaar, "U.S. military to 3D print its way out of supply chain woes," February 2022.
https://www.forbes.com/sites/carolynschwaar/2022/02/27/us-military-to-3d-print-its-way-out-of-supply-chain-woes/

[14] A. Jha, "The geopolitical implications of emerging military technologies for India," vol. 17, no. 6, Jan-Feb 2024.
https://www.defstrat.com/magazine articles/the-geopolitical-implications-of-emerging-military-technologies-for-india/#:~:text=The%20Geopolitical%20Implications%20of%20 Emerging%20Military%20 Technologies%20for%20India,-Sub%20Title%20%3A%20Emerging& text=This%20comprehensive%20 overview%20highlights%20the,adopting%20cutting%2Dedge%20military%20capabilities

[15] B. Koch and R. Schoonhoven (eds.), *Emerging Military Technologies: Ethical and Legal Perspectives.* Brill, 2022.

[16] B. R. Allen, *The Applied Ethics of Emerging Military and Security Technologies.* Routledge, 2015.

[17] Progressive Management, *21st Century Essential Guide to DARPA: Defense Advanced Research Projects Agency, Doing Business with DARPA, Overview of Mission, Management, Projects, DoD Future Military Technologies and Science.*Ebook.

[18] V. Gervais, *Emerging Technologies and the Future of Warfare.* TRENDS Research and Advisory, 2021.

[19] W. W. S. Wong, *Emerging Military Technologies: A Guide to the Issues.* ABC-CLIO, 2013.

[20] Sayler, *Emerging Military Technologies: Background and Issues for Congress.* Ebook, 2022.

[21] B. Koch and R. Schoonhoven (eds.), *Emerging Military Technologies Ethical and Legal Perspectives.* Brill, 2022.

[22] B. J. Strawser, G. R. Lucas Jr., and T. J. Demy, *Military Ethics and Emerging Technologies.* Taylor & Francis, 2016.

[23] R. M. O'Meara, *Emerging Military Technologies in the 21st Century: Assessing the Need for Governance.* Rutgers University, 2011.

[24] F. Barnaby and M.T. Borg (eds.), *Emerging Technologies and Military Doctrine*
A Political Assessment. United Kingdom: Palgrave Macmillan, 1986.

[25] National Research Council (US), *STAR 21: Strategic Technologies for the Army of the Twenty-First Century.* National Academies Press, 1992.

[26] R. M. O'Meara, *Governing Military Technologies in the 21st Century: Ethics and Operations.* Palgrave Macmillan, 2014.

CHAPTER 2

ARTIFICIAL INTELLIGENCE IN MILITARY

"The world's technocrats are yet to fully grasp and deploy the genius of Artificial Intelligence (AI) and the immensity of its powers to transform societies and elevate the quality of life of world citizens."
– Matthew N. O. Sadiku

2.1 INTRODUCTION

Technological development has become a rat race. New technologies that promise significant strategic advantages can upset balances or disrupt previously stable global governance arrangements. Artificial intelligence (AI) is one such critical technology. AI is an integral part of bringing technological advancements to the next level. It is among the many hot technologies that promise to change the face of warfare for years to come [1,2].

Artificial intelligence (AI) has done remarkable things such as defeating human experts at various games. AI is a technology that the military and defense world cannot ignore because the military cannot afford to miss out on the opportunities it brings. It has been described as the "third revolution" in warfare, after gunpowder and nuclear weapons. It has also been considered as the "fourth industrial revolution," which includes the Internet of things (IoT), nanotechnology, biotechnology, and robotics. It has become a critical part of modern warfare. It could cause drastic changes in hybrid warfare, which is a major concern for NATO [3].

Advances in AI, machine learning, and robotics are enabling new military capabilities that will have a disruptive impact on military strategies. The effects of these capabilities will be felt across the spectrum of military requirements – from intelligence, surveillance, marketing departments, and reconnaissance to offense/defense balances and even on to nuclear weapons systems themselves. Artificial intelligence and other emerging technologies will change the way war is fought. Whether it involves AI or not, war will always be violent, politically motivated, and composed of the same three elemental functions that new recruits learn in basic training: move, shoot, and communicate [4].

This chapter examines various applications of artificial intelligence in the military and defense. It begins by briefly reviewing the concept of AI. It covers what military is all about. It presents how AI is being incorporated in the military. It covers some applications of AI in the military. It highlights the benefits and challenges of military AI. It addresses how AI is being integrated in military forces around the globe.

It presents the future of military AI. The last section concludes with comments.

2.2 REVIEW ON ARTIFICIAL INTELLIGENCE

Artificial intelligence (AI) refers to computer systems that mimic human cognitive functions. It is a field of computer science that deals with intelligent machines. AI has long history which is actively and constantly changing and growing. The term "artificial intelligence" was first used at a Dartmouth College conference in 1956. The main goal of AI is to enable machines to perform complex tasks that typically require human intelligence [5]. In simple terms, AI attempts to clone human behavior. An important feature of AI technology is that is can be added to existing technologies. AI is now one of the most important global issues of the 21st century. It is poised to disrupt our world and change processes and developments in the fields of science, engineering, education, business, entertainment, agriculture, and military.

Artificial intelligence is an umbrella term that encompasses many different technologies. AI is not a single technology but a collection of techniques that enables computer systems to perform tasks that would otherwise require human intelligence [6].

The major disciplines in AI include:

- *Expert systems*
- *Fuzzy logic*
- *Neural networks*
- *Machine learning (ML)*
- *Deep learning*
- *Natural Language Processors (NLP)*
- *Robots*

These AI tools are illustrated in Figure 2.1. Each AI tool has its own advantages. Using a combination of these models, rather than a single model, is recommended. AI systems are designed to make decisions using real-time data. They have the ability to learn and adapt as they make decisions.

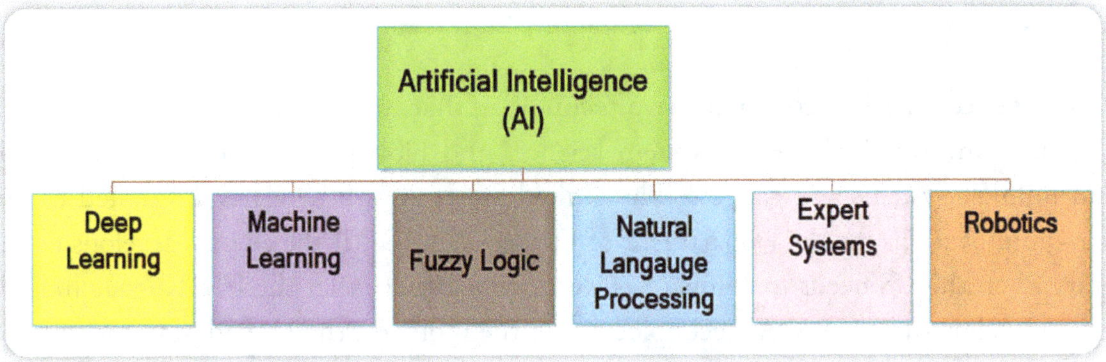

Figure 2.1 Branches of artificial intelligence.

AI has benefited many areas such as chemistry and medicine, where routine diagnoses can be initiated by AI-aided computers. Today, AI is integrated into our daily lives in several forms, such as personal assistants, automated mass transportation, aviation, computer gaming, facial recognition at passport control, voice recognition on virtual assistants, driverless cars, companion robots, wearables, etc. AI is emerging as the base technology for the military systems, where the future intelligent weapons are envisaged to transform the military operations. Now AI technologies are widely used in tactical warfare situations, such as target acquisition for missiles launched from drones. Military robots are usually employed within integrated systems that include video screens, sensors, grippers, and cameras [7,8].

2.3 THE MILITARY

The military, also known collectively as armed forces, is an armed and organized force primarily intended for warfare. It may consist of several branches such as army, navy, air force, space force, marines, and coast guard. The major job of the military is defined as defense of the state and its interests against external armed threats [9].

The US Department of Defense (DoD) was created in 1949. It comprises the Department of Army, Office of the Secretary of Defense, the Joint Chiefs of Staff, the military departments (Army, Navy, and Air Force, each under the authority of a civilian Secretary), 16 "defense agencies" which were created by the Secretary of Defense to perform particular functions, nine "Unified Commands" responsible for the conduct of military operations, civilians, contractors [10].

Battle is a physical activity and requires force. The victor could only employ the forces necessary to achieve victory through the advantage of foreknowledge. Winning a war also requires foresight, analysis, eyes and ears, and the development of strategies on how to win. That takes intelligence.

Today, nations have at their disposal information gathering systems such as radio, TV, satellites, ultramodern aircraft, human sources, cameras, and electronic devices. The United States government has devoted enormous resources to the creation and maintenance of a national intelligence system.

2.4 MILITARY AI

Artificial intelligence (AI) is a comprehensive technology that involves psychology, cognitive science, thinking science, information science, system science, and biological science. AI invades all major civilian and military systems and gadgets. The United States government has attempted to proliferate AI technology innovations for the Department of Defense (DOD). If the enemy develops better AI for their military, then the US needs to compete as well. Some believe that the US, Russia, and China are competing to develop and harness AI technologies. At the moment, the United States is the leading AI power, while China is emerging as a challenger. China is a strategic competitor with robust economic and technological capabilities. The DOD has created the Joint Artificial Intelligence Center with the intention of winning the AI battle and become the next great-power AI competitor.

The following key findings summarize a report military applications of AI [11]:

- A steady increase in the integration of AI in military systems is likely
- The United States faces significant international competition in military AI
- The development of military AI presents a range of risks that need to be addressed
- The US public generally supports continued investment in military AI

Figure 2.2 US ground troops patrol while robots carry their equipment [12].

Figure 2.2 shows a typical use of artificial intelligence in the military [12]. There are some problems with applying AI tools in the military and defense. These include [13]:

- Integrity of operation is of paramount importance
- Operation must often be in real time (millisecond responsiveness)
- It must be flexible in the face of changing circumstances
- It must be applicable in a domain in which even its most senior "practitioners" are in fact comfortable.

2.5 APPLICATIONS OF MILITARY AI

Artificial intelligence has the capability to help a decision-maker make better, more informed decisions. Militaries and defense organizations can use AI for autonomous weapons, autonomous vehicles, surveillance, cybersecurity, military intelligence, homeland security, logistics and transportation, military intelligence, and war planning. These applications are discussed as follows [14,15]:

Figure 2.3 Killer robots in wartime [16].

- *Autonomous Weapons:* Defense forces around the world are embedding AI into weapons and other systems used on land, naval, airborne, and space platforms. AI-based systems have enabled the development of efficient warfare systems, which are less reliant on human input. AI is also expected to empower autonomous and high-speed weapons to carry out attacks. US ground troops patrol while robots carry their equipment and drones serve as spotters. Figure 2.3 shows killer robots in wartime [16]. Military robots are better suited than humans for dull, dirty, repetitive, or dangerous tasks or missions. We should keep in mind that the public debate over the military use of AI mainly revolves around autonomous weapons systems.

- *Autonomous Vehicles:* AI is enabling autonomous systems to conduct missions, silent operations, automating tasks, and making better, quicker decisions than humans. An autonomous vehicle can operate with less regard for other drivers if its mission means saving the lives of one or more operators. It can drive itself using machine vision, creating a convoy. Boeing has offered autonomous drones and aircraft to militaries today and is currently designing autonomous submarines. Lockheed Martin has offered many AI-based solutions to the US military. Figure 2.4 illustrates an AI-powered autonomous armored vehicle [17].

Figure 2.4 AI-powered autonomous armored vehicles [17].

- *Weapons Targeting:* Targeting systems need to be accurate and quick to lock on targets. A human is capable of identifying an enemy vehicle, deciding a weapon system to employ against it, and then engaging the target. Today, autonomous weapon platforms use computer vision to identify and track targets. AI can be used for weapon targeting. This requires training the AI on what exactly a strategic target is worth focusing its firepower on and alerting the operator if necessary.

- *Surveillance:* Militaries around the world gather massive surveillance data a day from various sources, such as phone cameras, video surveillance, UAVs, and satellites. AI could be of help in the important task of processing the data for strategic information. The US DOD currently employs machine learning and computer vision software for surveillance operations.

- *Homeland Security:* One core capability of AI is predictive analytics, which is basically identifying patterns within a data set and then predict that trend will occur again. Predictive analytics models are currently being used in homeland security. Predictive analytics software can be used to give a prediction of possible suspects of a crime based on various environmental factors and past criminal record data.

- *Cybersecurity:* Military systems are vulnerable to cyber attacks. To avoid the high level of risk associated with cyber attacks, leaked government intelligence, and data breaches in military and defense networks, cybersecurity seems to be a high priority for the military. AI has the capability to play a vital role in preventative measures for the military. Some AI vendors use machine learning to offer security products that can identify and predict threats before they can affect the networks.

- *Logistics & Transportation:* Logistics (which is essentially the ability to supply forces with food, fuel, and replacements) has traditionally been the limiting factor in war. Military logistics is one area where AI could make a great impact. The effective transportation of goods, ammunition, armaments, and troops is an essential component of successful military operations. AI is expected to play a crucial role in military logistics and transport. Integrating AI with military transportation can help lower transportation costs and reduce human operational efforts. Military operators performing logistic support runs account for a minimum of 50% of the casualties while at war. AI is capable of allowing more efficient, data-backed logistics and maintenance of military equipment.

- *Battlefield Healthcare:* AI can be integrated with Robotic Surgical Systems (RSS) and Robotic Ground Platforms (RGPs) to provide remote surgical support in war zones. Under difficult conditions, systems equipped with AI can mine soldiers' medical records and assist in complex diagnosis.

- *Military Intelligence:* Modern warfare requires an integration of military and intelligence forces. Military intelligence is crucial and central to planning a victorious campaign. Intelligence is a conscious and necessary task assigned by leadership. Before the commander could determine how to employ his forces, he first has to know whether he can attack and where he should attack. Military intelligence is a military branch that uses information collection and analysis approaches to provide guidance and direction to assist commanders in their decisions. As an academic field, military intelligence is multidisciplinary area that combines language, political theory, economics, sociology, and psychology [18]. AI may be particularly useful for intelligence because of the proliferation of sensors and the availability of large data sets. The speed and precision of AI-enabled intelligence analysis can provide US forces an operational advantage against adversaries that do not possess similar capabilities.

- *Central Intelligence Agency* (CIA): AI capabilities in the CIA include discovering threats and thwarting planned attacks, neutralizing cyber attacks that come in through email, surveying areas via satellite, identifying and predicting social unrest in a region. The CIA finds modern innovations in AI useful for security and intelligence purposes [19].

- *War planning:* This is an area that desperately needs AI technologies. War plans are usually based on both assumptions and facts. As assumptions and facts change, the plan too changes. The plan may be based on units whose availability or mission changes. Using AI technologies, the plan could be automatically modified so that it is more than just shelfware [20].

These applications are simply a taste of what is ultimately possible. Other potential applications of AI in the military include shooting down drones, aiming tank guns, coordinating resupply, planning artillery barrages, blending sensor feeds, stitching different sensor feeds together into a coherent picture, analyzing how terrain blocks units' fields of fire, war games, combat automation in so-called manned-unmanned operations, and warning commanders where there are blind spots in their defenses.

2.6 BENEFITS

The military benefits immensely from AI technology. AI has many application areas where it will enhance productivity, reduce user workload, and operate more quickly than humans. The modern uses of AI in military are not limited to the battlefields. AI can help reducing the risk of life loss in wars. AI can be used for training systems [21]. Some AI applications will change many aspects of the global economy, security, communications, and transportation by altering how humans work, communicate, think, and decide. It improves self-control, self-regulation, and self-actuation of combat systems due to its inherent computing, and decision-making capabilities. .AI can be used to optimize communications in controlling how data and bandwidth are used effectively. As the use of AI grows, biases and discrimination inherent in AI will gradually disappear.

The following four benefits of AI in the military are changing the world of defense and national security [22]:

- *Autonomy:* Autonomous machines can be more efficient than regular soldiers. They are less bin cost about ten times compared with the cost of human soldiers. AI has the new capability to operate autonomous weapons at the miniaturized level. It increases the performance of warfare systems while minimizing the need for maintenance. It can automatically monitor weapons systems, mobile devices, and aircraft, which are vulnerable to cyber attacks. Autonomous armaments can accurately find and kill enemies on the battlefield. More automation is possible in the future. As technology moves beyond automation, autonomy and autonomous systems bring efficiencies to bear in several sectors.

- *Decision Making:* AI is critical for giving soldiers the ability to make informed decisions. AI systems will be used to identify and classify threats, prioritize targets, and show the location of friendly troops and safe distances around them. In the future, all combat decisions (such as targets and how much to fire to minimize collateral damage) could be made by robots, with humans monitoring the battlefield situation from a central command. Humans are better at making high-level decisions, while AI-enabled systems can process complicated things at high speed.

- *Machine Accuracy:* Machine accuracy is better than our decision-making error rate in life-or-death situations. In the future, machine accuracy at making combat-kill decisions will surpass human accuracy. The accuracy and precision of today's weapons are steadily forcing contemporary battlefields to empty of human combatants. Ships will have fewer crew members as the AI programs will do more. AI based warfare is rated superior to traditional warfare both in tactical and strategic standpoints.

- *Useful in Space:* The military AI has a place in space. In case of any Lunar Moon War, robots and drones will be sent first. Space Force cannot muster soldiers into rockets fast enough compared to launching remote AI drones and robots.

2.7 CHALLENGES

Some consider the term "artificial intelligence" as an oxymoron since it is regarded as the capability of a machine to imitate intelligent human behavior. The bar for what is considered "intelligent" keeps rising higher. Research shows that under adversarial conditions, AI systems can easily be fooled, resulting in wrong decisions. Many critics warn that AI may someday evolve beyond submission to their human controllers.

There have been proposals to ban or regulate the employment of autonomous weapons in a military operations. Moral objections to AI by some US citizens may slow new development by the DOD. Other challenges include [23].

- *Multi-tasking*: One of the weakness with AI-based systems is their inability to multi-task. A human can identify an enemy vehicle, decide which weapon to use, and then engage the target. This simple set of tasks is currently impossible to accomplish by an AI system. Most AI systems today are designed to perform a single task and they do not adapt well to new environments and new tasks as humans.

- *Full Autonomy:* Risks associated with military AI will require human operators to maintain positive control in its employment. Placing vulnerable AI systems in contested domains and making them responsible for critical decisions may lead to disaster. AI systems cannot be autonomous at this time; humans must be responsible for key decisions. The Human Rights Watch has advocated for the prohibition of fully autonomous AI-base system capable of making lethal decisions.

- *Hacking:* A machine can be hacked in ways a human cannot. The offensive use of AI malware is a major concern. Satellites in space, especially LEO satellites, are likely to remain highly vulnerable to nuclear attack.

- *Competition:* The desire to build new weapons for impending future conflicts has triggered an unhealthy arms race between the US and its competitors Russia and China. Efforts should be made to keep fast-paced advances in machine learning from sparking a worldwide AI arms race that poses a new existential risk to humanity. If autonomous machines supported by one country target and kill humans, other countries can follow suit, resulting in destabilizing global arms races. International competition in the development of military AI could lead to World War III. Perhaps AI based warfare would the final Armageddon.

- *Potential Abuses:* The introduction of AI technology needs oversight to prevent potential abuses and unintended consequences. AI and cyberspace could cause drastic changes in hybrid warfare, which is a major concern for NATO. A small nation can develop effective AI-based weapons without the industrial might needed to research and produce potent designs that will give it the edge needed to win a war.

- *Ethical Implications:* Ethical risks are important from a humanitarian viewpoint. Some thoughtful individuals have expressed serious, legitimate reservations about the legal and ethical implications of using AI in war or even to enhance security in peacetime. The use of lethal autonomous weapon systems (LAWS) raises some basic ethical and legal questions on human control. Governments and military leaders should understand the ethical and legal implications of employing the weapons. The ethical dictum of eye for an eye and tooth for a tooth would make the world blind and toothless; whereas an AI that truly implements the reaction of showing the other cheek would make the world utopia of peace.

2.8 GLOBAL AI IN MILITARY

The promise of AI (automation, informed decision making, self-control, self-regulation, and self-actuation of combat systems, etc.) is driving militaries around the world to accelerate research and development. Defense forces across the globe are seeking to gain an edge over their adversaries by integrating AI innovations into their arsenals. Many nations are developing AI for their policy guidance and strategic planning. They are increasingly deploying AI technology into weapons and other defense systems that are used on airborne, land, naval, and space platforms. More than thirty nations and international organizations have strategies and initiatives for AI. Various organizations such as NATO help spread knowledge, create awareness, stimulate research and development on AI technology. All NATO member states need to be involved in preparing for the transition to an AI-powered, highly interconnected world.

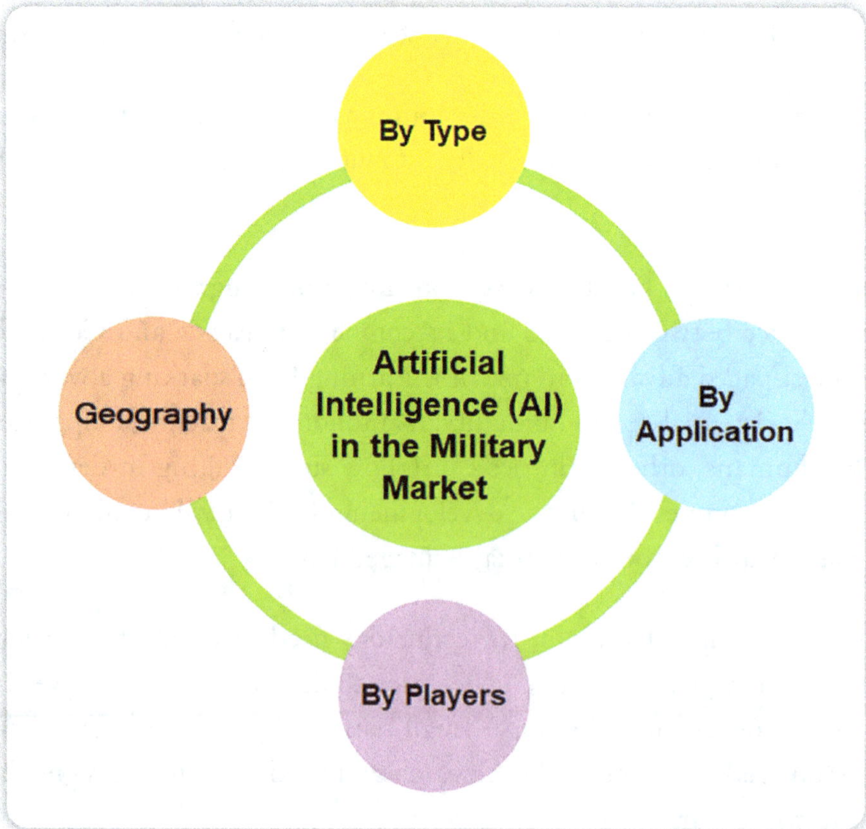

Figure 2.5 Segmentation of the global artificial intelligence in the military market.

The AI in Military market includes major players such as BAE Systems (UK), Northrop Grumman (US), Raytheon Technologies (US), Lockheed Martin (US), Thales Group (US), L3 Harris Technologies (US), Rafael Advanced Defense Systems (Israel), and IBM (US), etc. These players have spread their business across various countries including North America, Europe, Asia Pacific, Middle East & Africa, and Latin America [24]. Figure 2.5 shows the segmentation of the global AI in the military market. The development of military AI is accelerating due to seven key players: the United States, China, Russia, the United Kingdom, France, Israel, and South Korea. We typically consider the following countries and their interest in integrating AI in their military forces [25,26].

- *United States*: The US is recognized as one of the key manufacturers, exporters, and users of AI systems worldwide primarily due to the presence of leading tech companies such as Microsoft, Google, IBM, Northrop Grumman, and others willing to invest in the technology. The primary goal of the US military's AI strategy is to improve the readiness of troops and increase lethality. The US intends to increase its spending on AI in the military to gain a competitive edge over other nations. AI supports and protects US servicemembers, safeguards US citizens, defends US allies, and improves the affordability, effectiveness, and speed of US military operations. The US military sees many benefits in pairing humans with intelligent technologies. The National Defense Authorization Act (NDAA) established a Joint Artificial Intelligence Center (JAIC) under the DoD to oversee about 600 active AI projects [27]. AI has been used to deliver military training in the United States. The Pentagon is spreading AI-powered technologies faster throughout the services. Countries like Russia and China are no longer looking to achieve parity with the US; they want to surpass it by researching heavily into the weapons of the future. The world will be safer and more peaceful with US leadership in AI.

- *Russia*: Russia has declared a new frontier for military research. In 2017, President Vladimir Putin mentioned that whoever became the leader in the sphere of AI would "become the ruler of the world." To back that up, the same year Russia's Military-Industrial Committee approved the integration of AI into 30 percent of the country's armed forces by 2030. Russia has stated that the debate around lethal autonomous weapons should not ignore their potential benefits. There should be an increasing cooperation between military and civilian scientists in developing AI technology. The binary Russian–US nuclear rivalry, legacy of the old Russian–US confrontation, is being gradually replaced by regional nuclear rivalries.

- *China*: China has stated that a major focus of research and development is how to win at "intelligent(ised) warfare." Current areas of focus include STEM education, AI-enabled radar, robotic ships, smarter cruise and hypersonic missiles, all areas of research that other nations are focusing on. Everything from submarines to satellites, tanks to jets, destroyers to drones, are AI connected by China. China intends to be the global leader of AI by 2030. Beijing regards AI as a critical component to its future military and industrial power. The Chinese government as well as Chinese companies have invested heavily in expanding their computing power and semiconductor capabilities to narrow the gap with the West. The government has been researching air, ground, surface, and undersea autonomous unmanned vehicles (AUVs), which can employ AI to perform autonomous guidance, target acquisition, and attack execution. There have been calls from within the Chinese government to avoid an AI arms race, which could lead to a World War III [28].

- *European Union:* In 2019, some EU member states called for greater collaboration between EU members on the military AI. They find concrete ways for the EU and its member states to work towards common principles and best practices for the responsible military use of AI. [29]. For example, France understands the autonomy of LAWS (lethal autonomous weapon systems) as total, with no form of human supervision.

- *United Kingdom:* The UK is of the position that an autonomous system is capable of understanding higher level intent and direction. It stated that the current lack of consensus on key themes counts against any legal prohibition. The Ministry of Defense (MoD) is pursuing modernization in AI and related technologies. The MoD has various programs related to AI and autonomy [30].

- *India:* India is emerging as the hub for "Digital Skills." The necessity of AI has been realized in India. From information to decision making to direct destruction of military capabilities, AI will be used. However, the use of AI in the Indian military is expected to begin in the near future. It will take around 3-4 years before the AI tool is used in the Indian military. It is intended that every army personnel will be having the tools that are integrated with AI. The defense ministry had set up a multi-stakeholder task force for Strategic Implementation of Artificial Intelligence and Defense [31].

2.9 FUTURE OF MILITARY AI

Modern warfare is based on unprecedented connectivity of military systems. Artificial intelligence will certainly play a major role in future military applications. In the future, AI systems that can be trained to learn and think independently will likely dominate the field of AI. Here we will consider the future of warfare, the future of technology, and the future policy on AI.

- *Future of Warfare:* In spite of the subjectivity of predicting the future of warfare, one can identify the following five overarching trends that will help shape the who, what, where, and how of warfare in the decades to come [32]:

  - Trend 1: the competition for regional hegemony will increase
  - Trend 2: defending ground will become more challenging
  - Trend 3: the American qualitative and quantitative military edge will decline
  - Trend 4: the lines between war and peace will continue to blur
  - Trend 5: the war on terrorism will continue

- *Future of Technology*: It is difficult to predict the exact impact of AI-enabled technologies. However, it is clear that AI is poised to transform warfare in the near future. AI-powered platforms are the future of any battlefield. AI will support armed forces in collecting, categorizing, and analyzing data more quickly and efficiently than is currently possible. As AI technology improves, a constellation of military devices could be made largely autonomous. Unmanned underwater vehicles (UUVs) could be widely deployed in times of crisis. AI and robotics will continue to play a central, decisive role in future battles or warfare. Looking toward the future, one can imagine a fundamental change in the character of war. It is highly likely we will eventually see fully autonomous weapons on the battlefield.

- *Future Policy:* The future policy on AI and national security involves preserving US technological leadership, supporting peaceful and commercial use, and mitigating catastrophic risk. By looking at four prior cases of transformative military technology—nuclear, aerospace, cyber, and biotech—we develop lessons learned and recommendations for national security policy toward AI [33]:

    - Lesson #1: Radical technological change begets radical government policy ideas
    - Lesson #2: Arms races are sometimes unavoidable, but they can be managed
    - Lesson #3: Government must both promote and restrain commercial activity
    - Lesson #4: Government must formalize goals for technology safety and provide adequate resources
    - Lesson #5: As technology changes, so does the United States' National Interest

Future progress in AI has the potential to be a transformative national security.

2.10 CONCLUSION

Artificial intelligence is a rapidly growing branch of computer science which requires computer programming. It is a rapidly developing capability and AI models are improving daily. The use of AI in everyday life increases. AI will change how wars are planned and fought. It also has many military application areas where it will enhance productivity, reduce user workload, and operate more quickly than humans. It has the capability to gather and quickly synthesize information from many sources to produce highly accurate estimates of locations for submarines, or land-based mobile launchers. AI technologies should be used to supplement rather than replace human ingenuity, creativity, and judgement.

Current military doctrine assigns command and control responsibilities to humans, not to machines. Advances on AI will determine their future strategic effectiveness in military matters, as well as their performance, competitiveness, and ability to deter adversaries. These advances in hardware are what enable the "internet of things," and what will become the internet of battlefield things. Artificial intelligence will have immense impact on national and international security. For more for information about AI in the military, one should consult the books in [29,34-39].

REFERENCES

[1] "Artificial intelligence in military application information technology essay," *Information Technology*, January 1970.

[2] M. M. Maas, "How viable is international arms control for military artificial intelligence? Three lessons from nuclear weapons," *Contemporary Security Policy,* vol. 40, no. 3, 2019, pp. 285-311.

[3] M. N. O. Sadiku, S. R. Nelatury, and S. M. Musa, "Artificial intelligence in military," *Journal of Scientific and Engineering Research,* vol. 8, no. 1, 2021, pp. 106-112.

[4] C. Brose, "The new revolution in military affairs," *Foreign Affairs,* vol. 98, no. 3, May/June 2019, pp. 122-128,130-134.

[5] S. Greengard, "What is artificial intelligence?" May 2019, https://www.datamation.com/artificial-intelligence/what-is-artificial-intelligence.html

[6] https://in.pinterest.com/pin/828662400161409072/

[7]" Chapter 12 - Specialized machine learning," https://sanjeevkatariya.github.io/ai/machinelearning/chapter-6/index.html

[8] Y. Mintz and R. Brodie, "Introduction to artificial intelligence in medicine," *Minimally Invasive Therapy & Allied Technologies,* vol. 28, no. 2, 2019, pp. 73-81.

[9] "Military," *Wikipedia,* the free encyclopedia https://en.wikipedia.org/wiki/Military

[10] "Military intelligence," https://fas.org/irp/offdocs/int014.html

[11] F. E. Morgan et al., "Military applications of artificial intelligence ethical concerns in an uncertain world," https://www.rand.org/pubs/research reports/RR3139-1.html

[12] "Artificial intelligence and the military," https://www.rand.org/blog/2017/09/artificial-intelligence-and-the-military.html

[13] "Artificial intelligence in defence: Wanted and unwanted research." *IEE Colloquium on Strategic Industrial Issues in AI in Engineering,* January 1991.

[14] M. Roth, "Artificial intelligence in the military – An overview of capabilities," February 2019, https://emerj.com/ai-sector-overviews/artificial-intelligence-in-the-military-an-overview-of-capabilities/

[15] T. Singh and A. Gulhane, "8 key military applications for artificial intelligence in 2018," October 2018, https://blog.marketresearch.com/8-key-military-applications-for-artificial-intelligence-in-2018

[16] "Killer robots in wartime: Could they be more deadly than humans?" April 2017, https://towardfreedom.org/story/archives/globalism/killer-robots-wartime-deadly-humans/

[17] Y. Lappin, "Israel seeks to change the face of the battlefield with AI-powered autonomous armored vehicles," https://www.jns.org/israel-seeks-to-change-the-face-of-the-battlefield-with-ai-powered-autonomous-armored-vehicles/

[18] M. N. O. Sadiku, O. D. Olaleye, A. Ajayi-Majebi, and S. M. Musa, "Military intelligence: A primer," *International Journal of Trend in Research and Development*, vol. 7, no. 3, 2020, pp. 298-302.

[19] M. Roth, "Artificial intelligence at the CIA – Current applications," November 2019, https://emerj.com/ai-sector-overviews/artificial-intelligence-at-the-cia-current-applications/

[20] K. J. Carlson, "The military application of artificial intelligence," Unknown Source.

[21] H. Soffar, "Military artificial intelligence (military robots) advantages, disadvantages & applications," August 3029, https://www.online-sciences.com/robotics/military-artificial-intelligence-military-robots-advantages-disadvantages-applications/

[22] G. Cooke, " Magic bullets: The future of artificial intelligence in weapons systems," June 2019, https://www.army.mil/article/223026/magic bullets the future of artificial intelligence in weapons systems#:~:text=Magic%20Bullets%3A%20The%20Future%20of%20Artificial%20 Intelligence%20in%20Weapons%20Systems,-By%20Dr.&text=We%20live%20in%20an%20era,today's %20widely%20adopted%20consumer%20product.&text=And%20they%20bring%20with%20 them,than%20any%20science%20fiction%20story.

[23] P. Maxwell, "Artificial intelligence is the future of warfare (just not in the way you think), April 2020, https://mwi.usma.edu/artificial-intelligence-future-warfare-just-not-way-think/

[24] "Global Artificial Intelligence in Military Market (2020 to 2025) - Incorporation of quantum computing in AI presents opportunities," March 2021, https://www.businesswire.com/news/home/20210323005739/en/Global-Artificial-Intelligence-in-Military-Market-2020-to-2025---Incorporation-of-Quantum-Computing-in-AI-Presents-Opportunities---ResearchAndMarkets.com

[25] "Global military artificial intelligence (AI) market size by type, by applications, by geographic scope and forecast," https://www.verifiedmarketresearch.com/wp-content/uploads/2020/09/Slide2-2020-10-09T025303.710.jpg

[26] A. Gatopoulos, "Project Force: AI and the military – a friend or foe?" March 2021, https://www.aljazeera.com/features/2021/3/28/friend-or-foe-artificial-intelligence-and-the-military

[27] "The future of military (artificial) intelligence," April 2021, https://www.designnews.com/electronics-test/future-military-artificial-intelligence

[28] Y. Tadjdeh, "China threatens U.S. Primacy in artificial intelligence (UPDATED)," October 2020, https://www.nationaldefensemagazine.org/articles/2020/10/30/china-threatens-us-primacy-in-artificial-intelligence

[29] V. Boulanin et al., *Responsible Military Use of Artificial Intelligence: Can the European Union Lead the Way in Developing Best Practice?* SIPRI, 2020.

[30] K. Gronlund, " May 2019, https://futureoflife.org/2019/05/09/state-of-ai/

[31] H. Siddiqui, "Future warfare: Is Indian army ready for the use of artificial intelligence and smart technologies?" December 2020,

https://www.financialexpress.com/defence/futre-warfare-is-indian-army-ready-for-the-use-of-artificial-intelligence-and-smart-technologies/2145585/

[32] J. D. Winkler et al., "Reflections on the future of warfare and implications for personnel policies of the U .S. Department of Defense," https://www.rand.org/pubs/perspectives/PE324.html

[33] G. Allen and T. Chan, "Artificial intelligence and national security," July 2017, https://www.belfercenter.org/publication/artificial-intelligence-and-national-security

[34] Artificial Intelligence in Military. Independently Published, 2020.

[35] P. J. Springer, *Military Robots and Drones: A Reference Handbook*. Santa Barbara, CA: ABC-CLIO, 2013.

[36] J. I. Walsh and M. Schulzke. *Drones and Support for the Use of Force*. Ann Arbor, MI: University of Michigan Press, 202.

[37] G. Allen and T. Chan, *Artificial Intelligence and National Security*. Abebooks, 2017.

[38] K. Payne, *Strategy, Evolution, and War: From Apes to Artificial Intelligence*. Georgetown University Press, 2018

[39] M. Cummings, *Artificial Intelligence and the Future Of Warfare*. Chatham House for the Royal Institute of International Affairs, 2017.

CHAPTER 3

ROBOTICS IN THE MILITARY

"Robots have already surpassed human beings in calculation and memory, but I have no doubt that the time will come when they will surpass in wisdom as well."
– Masayoshi Son

3.1 INTRODUCTION

Robotics constitutes one of the most exciting fields of technology today. It is the discipline of designing and constructing intelligent machines, called robots. A robot is an autonomous mechanical device that is designed to sense its environment, carry out computations to make decisions, and perform actions like humans in the real world. Popular interest in robotics has increased in recent years. Robots are becoming more and more common in our society and more integrated into our lives. This is due to the fact that they are becoming smarter, smaller, cheaper, faster, more flexible, and more autonomous than ever before. Robotics technology has been implemented in a variety of fields including manufacturing, medicine, elderly care, rehabilitation, education, agriculture, home appliances, search and rescue, car industry, defense, and more.

Robotics constitutes one of the most exciting fields of technology today, presenting new applications for autonomous systems that can impact everyday life. Today, there are robots that can autonomously sense, reason, plan, act, move, communicate, and collaborate with other robots. The robotics revolution is going to change us as humans [1].

Robotics is the discipline of designing and constructing machines, called robots. A robot is an autonomous mechanical device that is designed to sense its environment, carry out computations to make decisions, and perform actions like humans in the real world. It is a system that contains sensors, control systems, power supplies, and software, all working together to perform a task. Robotics is a relatively young field with highly ambitious goals. It is producing a huge range of devices, from autonomous vacuum cleaners to military drones [2].

Technology is transforming how humans and machines work together. Advancements in technology have reshaped the landscape of military operations, with robotics emerging as a transformative force. The integration of robotics in the military has ushered in a new era of efficiency, precision, and adaptability on and off the battlefield. The use of robots in warfare is being researched as a possible future means of fighting wars.

A military robot is an independent robot or remote-controlled mobile robot specially intended for military applications. Military robots save military lives by using these robots in applications that could be dangerous for human personnel. Some believe the future of modern warfare will be fought by automated weapons systems [3].

This chapter examines the various uses of robots in the military. It begins with discussing what robots are all about. It discusses military robots and their various types. It presents some applications of military robots. It addresses military robots around the world. It highlights the benefits and challenges of military robots. It concludes with comments.

3.2 WHAT ARE ROBOTS?

The word "robot" was coined by Czechriter Karel Čapek in his play in 1920. Isaac Asimov coined the term "robotics" in 1942 and came up with three rules to guide the behavior of robots and later added the zeroth law [4]:

- Law 0: A robot may not injure humanity or through inaction, allow humanity to come to harm.
- Law 1: Robots must never harm human beings,
- Law 2: Robots must follow instructions from humans without violating rule 1,
- Law3: Robots must protect themselves without violating the other rules.

Robots are becoming increasingly prevalent in almost every industry, from healthcare to manufacturing. Figure 3.1 indicates that robotics is one of the branches of artificial intelligence.

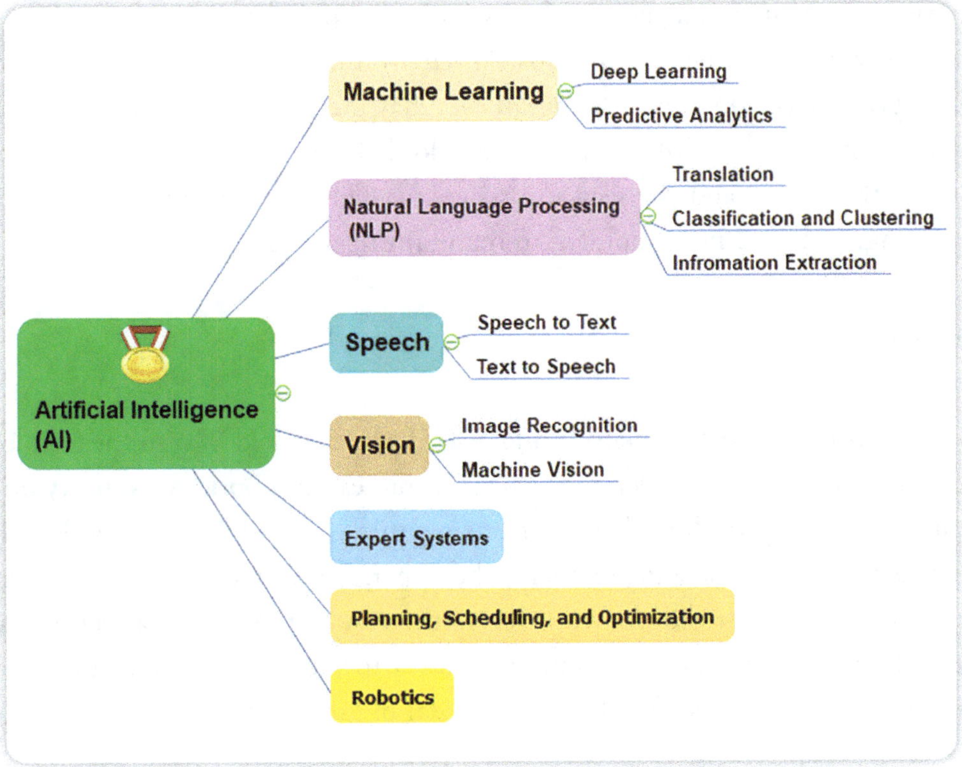

Figure 3.1 Robotics is one of the branches of artificial intelligence.

Although there are many types of robots designed for different environments and for different purposes/applications, they all share four basic similarities [5]:

(1) All robots have some form of mechanical construction designed to achieve a particular task;
(2) They have electrical components which power and control the machinery;
(3) All robots must be able to sense its surroundings; a robot may have light sensors (eyes), touch and pressure sensors (hands), chemical sensors (nose), hearing and sonar sensors (ears), etc.
(4) All robots contain some level of computer programming code.

Programs are the core essence of a robot since they provide intelligence. There are three different types of robotic programs: remote control, artificial intelligence, and hybrid. Some robots are programmed to faithfully carry out specific actions over and over again (repetitive actions) without variation and with a high degree of accuracy.

Robotics is an interdisciplinary field that involves the design, construction, operation, and use of robots. It is a branch of engineering and computer sciences that includes the design and use of machines that are capable of performing programmed tasks without human involvement. The field develops machines that can efficiently carry out various tasks, can automate tasks, and do various jobs that a human might not be able to do. Robots could someday be our drivers, companions, collaborators, teachers, specialists, and exploration pioneers.

The goal of robotics is to create intelligent machines (called robots) that behave and think like humans. Robots were originally intended for use in industrial environments to replace humans in tedious and repetitive tasks. Today, robots help human beings in everyday life. They are regarded as intelligent agents that can perform actions similar to what humans can do. It is not just on the ground, military robots have been taking to the skies—and the seas and space, too. They are being used for everything from maintaining, painting, and sanding equipment to maintaining aircraft.

3.3 MILIRARY ROBOTS

Broadly defined, military robots date back to World War II and the Cold War in the form of the German Goliath tracked mines and the Soviet teletanks. Military interest in robotics was spotty during the Cold War, with inventors repeatedly finding that what was technically possible mattered less than what was bureaucratically feasible. In spite of some setbacks, the American military robotics community did not waver in its belief in the usefulness of its work. Over the rest of the 1990s, as sensors and computer processors improved, unmanned systems became ever more capable. With a change in military mentality, money, and use, the groundwork was finally laid for a real military robotics industry.

Many of the military robots in development today can perform highly skilled tasks with remarkable precision. The majority of them are tele-operated and not equipped with weapons; they are used for reconnaissance, surveillance, sniper detection, neutralizing explosive devices, etc. Military robots are used for detecting and disarming bombs, underwater scouting, cargo transportation, and more. The vast majority of military robots are being used behind the scenes to keep everything up and running. American soldiers have been known to name the robots that serve alongside them. These names are often in honor of human friends, family, celebrities, pets, etc.

3.4 TYPES OF MILIRARY ROBOTS

The military and defense organizations have developed different kinds of robots to drive efficiencies, protect the lives of soldiers, traverse hazardous environments, respond to emergencies, and improve mission accuracy. Examples of military robots include the following [6]:

1. *Armed Robots:* These robots are equipped with weapons to help the military eliminate threats in combat, improve their operations, and help in defense systems,

2. *Unmanned Ground Vehicles:* These robots are used for a variety of purposes, including bomb disposal, reconnaissance, communications, and patrolling in challenging terrains.

3. *Training Robots:* These military robots offer training solution that simulates live combat.

4. *GuardBot:* This spherical amphibious robot can roll on any terrain and is used for surveillance, security, and military missions.

5. *Transportation Robots:* This robotic technology can help soldiers transport different supplies like artillery, bombs, and other supplies. These are also called logistics robots.

6. *Search and Rescue Robots:* These military robots are designed to assist in search and rescue missions.

7. *Mine Clearance Robots:* These are military robots that are designed specifically to detect and clear land mines. An example of mine clearance robot is shown in Figure 3.2 [6].

Figure 3.2 A mine clearance robot [6].

8. *Firefighting Robots*: These military robots can assist in extinguishing fires that may occur during combat.

9. *Surveillance Robots*: These robots can assist the military to survey and spy on the enemy or potential threats.

10. *Defense Robots:* These are professional service robots that are deployed by the military in combat scenarios, protecting, and enabling soldiers in combat.

11. *Robotic Dog*: A robotic dog is a four-legged robot that moves across a sandy desert landscape. An example of a robotic dog is shown in Figure 3.3 [7].

Figure 3.3 A robotic dog [7].

12. *Robotic Mules:* Robotic mules could transform the logistical backbone of military operations, featuring all-terrain mobility to navigate through the harshest landscapes. Figure 3.4 shows some examples of robotic mules [8].

Figure 3.4 Examples of robotic mules [8].

13. *Military Drones*: These are flying robots that are primarily used for reconnaissance and scouting, able to venture into dangerous territory and retrieve intelligence.

14. *Service Robots:* Since the military is never sure of what they are going to encounter on a mission, service robots can be sent into dangerous situations to keep soldiers at a safe distance.

15. *Submarine Robots:* These are robots that operate in the sea domain. They are also called underwater robots. An example of underwater robot is shown in Figure 3.5 [9].

Figure 3.5 Underwater or submarine robot [9].

3.5 APPLICATIONS OF MILITARY ROBOTS

Military robots are used for a variety of purposes including defense, search and rescue, surveillance, security, and transportation. Today, robots are used for a wide range of tasks, from reconnaissance to combat. The military is using robotics in the following specific ways [10]:

1. *Military Transportation:* Transporting military equipment, such as bombs and artillery supplies, carrying soldiers to and from battlefields, and recovering injured soldiers or casualties are different roles that can be performed by a military transportation robot. These robots can drive logistics efficiencies, reduce the physical burden placed on soldiers, and navigate difficult terrains in adverse weather conditions. They are typically unmanned ground vehicle (UGV) robots that come equipped with wheels or legs.

2. *Surveillance:* Fitted with weapons, high-sensor cameras, and infrared vision, surveillance robots are critical to military operations. They enable the close monitoring of enemy territories to provide battlefield intelligence and, ultimately, tactical advantage.

   These robots negate the need for human scouts, and help safeguard the lives of all soldiers on a mission. For example, Guardbot is a spherical amphibious surveillance robot that can roll on any terrain, including snow, sand, and dirt. Though originally developed for missions to Mars, the robot has applications in surveillance, security, and military missions.

3. *Search and Rescue:* Casualties often occur in war zones because it is too difficult or too risky to extract injured soldiers. These military robots provide critical support in finding missing or captured personnel. They can often go where humans cannot, whether underwater, through floods, wildfires, or over mountains. Autonomous or remote-controlled search and rescue robots can drastically reduce emergency response time, accurately pinpointing the location of human life and immediately embarking on a rescue mission. These robots are especially useful in disaster relief such as supporting local search and rescue after a tsunami, earthquake, or other natural disasters. For example, a dog-like robot called Jueying X20 is able to navigate uneven terrains. It can carry important medical equipment such as oxygen tanks, making it invaluable in a search and rescue mission.

4. *Neutralizing:* There are robots that detect and remove landmines and sea mines and sophisticated robotic arms that neutralize suspicious objects or bombs without putting a single human life at risk. These robots are typically used for route clearing, checkpoints, and vehicle inspections.

5. *Drones:* Drones are perhaps the most well-known application of robotics in the military. These flying robots are either remotely controlled or programmed to fly autonomously through software-controlled flight plans. Military drones are often the first responders in an emergency situation. With the ability to hover over an area for an extended period, they can assess hazards, gather intelligence, and pinpoint lost or wounded soldiers. For example, US Air Force has developed the technology to equip autonomous drones with face recognition technology. These drones are intended for use by special operations forces to gather intelligence ahead of missions in foreign countries. Robot drones, mine detectors, and sensing devices are common on the battlefield but require direct control by humans. Figure 3.6 depicts a solder assembling a Ghost-X drone [11].

Figure 3.6 A solder assembling a drone [11].

6. *Firefighting:* Fires are commonplace in combat and they remain one of the biggest threats to shipboard life, but that is something the US Navy is hoping to change. In 2015, it unveiled SAFFiR (Shipboard Autonomous Firefighting Robot), a humanoid-type robot that can navigate a ship's narrow passageways and has enough battery power for 30 minutes of firefighting. It can move through black smoke, manipulate fire-suppressing equipment, and throw propelled extinguishing agent technologies grenades. These military robots can be linked up to a fire hydrant to help investigate a fire site as well as put out fires to save victims› and firefighters› lives.

7. *Autonomous Vehicles:* Remote-controlled or autonomous vehicles are the future of military transportation robots since they can advance into a war zone, or another hazardous situation, without putting additional human lives at risk. They are designed for various military applications, from transport to search and rescue and attack. The self-driving vehicles will aid in the automation of future US Army ground vehicles, designed to navigate complex terrain and provide the army with the ability to remotely operate vehicles during high-risk missions. There have been some developments towards developing autonomous fighter jets and bombers. However, military weapons are prevented from being fully autonomous; they require human input at certain intervention points to ensure that targets are not within restricted fire areas as defined by Geneva Conventions for the laws of war.

3.6 MILITARY ROBOTS AROUND THE WORLD

The United States is far from the only country interested in robotics capabilities. Robotics is one of the hot fields of modern age in which the nations are concentrating upon for military purposes in the state of war and peace. The global military robots market size is expected to reach $32.7 billion by 2030. However, the rapid development of disaster and battlefield-ready robots has left the United Nations and human rights groups uneasy. We consider how military robots are being used in the following nations.

- *United States:* When it comes to military use of robots, the US leads the way despite leadership changes and funding issues, and continues to do so in ethical ways, with international support. In the United States, the Department of Defense (DoD) is the largest customer for unmanned systems technologies. The US military is investing heavily in the RQ-1 Predator, which can be armed with air-to-ground missiles and remotely operated from a command center in reconnaissance roles. A single aircraft carrier costs billions of dollars, and the United States relies heavily on its ten aircraft carrier strike groups to project power around the globe. But as military robots match more capabilities found in nature, things might change. The U.S. military is now experimenting with sentient unmanned vehicles that literally think like their human creators.

- *China:* China is no stranger to robots. China and its connecting regions have a basic influence to play that will help the market with fostering the locale. The overall autonomous robots market predicts the creating responsibility of associations like Alibaba, Walmart, and Amazon, which encourages a creating necessity for AGVs for dispersion focus motorization on a drawn-out scale for the smooth conduction of the market exercises. The Chinese People's Liberation Army (PLA) is in possession of the small ground robot, which can traverse complicated terrains, accurately observe battlefield situations and provide ferocious firepower. Equipped with a machine gun, and observation and detection equipment including night vision devices, the robot can replace a human soldier in dangerous reconnaissance missions.

- *Russia:* Russia's defense leadership, military theorists, and military practitioners are showing keen interest in robotic military applications featuring varying degrees of autonomy in performing their tasks. Moscow's military campaigns against Ukraine and in Syria have become the testbeds of such applications.

- *India:* Today, India is one of the developing nations in the world. The country's defense plays an important role in uplifting the nations growth and strengthening the defense system is most important to protect the country and the people. Indian military aims to add an additional feature to defense mechanism which helps to fight against terrorism and antisocial activities. Their Defence Rover is a military based application robot that can used in critical conditions such as surgical strikes, secret sting operations, anti-violence, anti-terrorism, and during natural disasters. It is a man operated system. It is designed to combat terrorism and natural disasters. It is a highly adaptable, autonomous vehicle equipped with advanced sensors, communication systems, and specialized tools. The striking feature of this rover is that it is able to fly in air as a drone and also travel in water. The Defense Rover is shown in Figure 3.7 [12].

Figure 3.7 Indian Defense Rover [12].

3.7 BENEFITS

Robotics for military applications is a rapidly growing and evolving field that offers many benefits and challenges for the military and society. Perhaps the main reason robots are crucial to the military mission is their ability to protect human life. Robots are cheaper to make than training and caring for living, breathing human soldiers. Since military robots are lifeless technologies, they can perform challenging and even hazardous tasks that would be otherwise risky or impossible for humans. They help reduce labor-intensive tasks and reduce the risk of threats, accidents, injury, and death for their human counterparts. Other benefits of military robots include the following [13]:

1. *Impersonalizing War:* Unmanned systems have a profound effect on the impersonalization of battle. Robots do not have emotion or an instinct of survival and will not lash out in fear. They show no anger or recklessness because they are not programmed to. As AI improves, we will see computers that can carry out orders more efficiently and reliably after they are programmed to do so, and they will not think twice about their orders. The hope that technology will reduce the violence of war is a venerable one. Some analysts believe that robot warriors can help reduce the flow of blood and perhaps make war more moral. The war weapons do not just create greater physical distance, but also a different sort of psychological distance and disconnection.

2. *Cost Reduction:* Robots can reduce the costs and risks of military operations by replacing or supplementing human personnel, equipment, and vehicles. Robots will be much less expensive than hiring soldiers. However, in order to deploy autonomous killers, we must weigh the costs and benefits over many issues. Since the cost of human life is the largest cost any technology can accrue, replacing humans with robots is reasonable and ethical.

3. *Reducing Mistake:* Many wartime atrocities are not the result of deliberate policy or fits of anger; they are just human mistakes. Unmanned systems seem to offer several ways of reducing the mistakes and unintended costs of war. Their precision superior to what humans could marshal on their own can lessen the number of mistakes made. It is easy to see how collateral damage can be greatly reduced by robotic precision.

4. *Reduction of Risk*: One of the primary benefits of deploying robotics in military operations is the ability to reduce risks for human soldiers. The removal of risk allows decisions to be made in a more deliberate manner than normally possible. Soldiers describe how one of the toughest aspects of fighting in cities is how you have to burst into a building and, in a matter of milliseconds, who is an enemy and who is a civilian and shoot the ones that are a threat before they shoot you. A robot can only shoot at someone who shoots first, without endangering a soldier's life. Some feel that unmanned systems can remove the anger and emotion from the humans behind them. Figure 3.8 illustrates how robots can save wounded soldiers from the battlefield [14].

Figure 3.8 Robots can save wounded soldiers from the battlefield [14].

5. *Human in the Loop*: While military robots are useful for many applications, every implementation of military robots includes a human being (usually several). Robots can do a lot of different things, but humans are needed at every step, from building to maintaining to programming to overseeing. Even more worrisome, the concept of keeping human beings in the loop is already being eroded by policymakers and by the technology itself.

6. *Robotic Jobs*: Where there are robots, there are robotics jobs, and military robotics programs are no exception. When we talk about robotics jobs, we are usually speaking about robotics in the manufacturing industry. The military needs robotics workers just as much as the manufacturing industry, because they use robotics to maximize their effectiveness. Robots are primarily protecting people and improving the safety of personnel.

7. *Increased Trust:* The military is building trust through rigorous validation and verification of their autonomous capabilities to mitigate direct threats to the warfighter – ultimately, increasing mission readiness.

8. *Maximized Performance:* The deployment of autonomous systems help to maximize the operational effectiveness of a mission while ensuring the safety of the human-in-the-loop across air, land, sea, and space.

9. *Enhanced Precision:* Robotics in the military significantly enhance precision in various tasks, from reconnaissance to targeted strikes. Unmanned systems can be equipped with advanced sensors and targeting technologies, allowing for surgical precision in operations while minimizing collateral damage. This precision contributes to the overall reduction of civilian casualties and infrastructure damage.

10. *24/7 Surveillance and Monitoring:* Unmanned systems, including drones and surveillance robots, provide continuous monitoring capabilities. This 24/7 surveillance enhances situational awareness, allowing military forces to respond quickly to emerging threats.

11. *Logistical Efficiency*: Autonomous ground vehicles and drones can be employed for logistical tasks, including supply transport and resupply operations. This not only increases the efficiency of logistical processes but also reduces the burden on human personnel, freeing them up for more strategic and specialized tasks.

12. *Flexibility:* Robots can improve the efficiency and effectiveness of military operations by increasing speed, accuracy, endurance, and flexibility. Robotics in the military offer a high degree of adaptability and flexibility. Unmanned systems can be rapidly deployed and reconfigured for various missions, allowing military forces to respond dynamically to changing circumstances.

13. *Cost-effectiveness:* The Department of Defense (DOD) spends millions of dollars on training soldiers to wage war. Training one soldier costs between $50,000 to $100,000, with an annual cost of at least $100,000 to maintain the soldier's health, training, and other benefits like salary and housing. In contrast, robots require about the same initial cost but much less in maintenance and storage. Using robots in the U.S. Army would reduce the cost of recruiting, both from the monetary and behavioral health impacts.

3.8 CHALLENGES

There are many challenges ahead to integrate robots and soldiers on the battlefield. The robotics industry is constantly changing and evolving. Humans do not know what the ultimate technological performance limit for autonomous robotics is. Robots can create technical and operational difficulties for the military, such as the reliability and security of the systems, the interoperability and integration with other systems, and the training and maintenance of the operators. Beyond technical challenges, the Army must convince Congress to alter the procurement process so the service can acquire or adapt capabilities within broader funding lines. Other challenges facing military robots include the following:

1. *Ethics*: Robotics for military applications is a rapidly growing and evolving field that offers many benefits and challenges for the military and society. The use of robots in warfare raises ethical and moral questions. In the heat of battle, fear, anger, and vengefulness can cause even the most trained soldiers to commit war crimes that violate ethical standards laid down by Geneva and other international conventions. As military robots gain more and more autonomy, the ethical questions involved will become even more complex. Autonomous machines are created not to be "truly 'ethical' robots", yet ones that comply with the laws of war (LOW) and rules of engagement (ROE). This is the main reason military lawyers are so concerned about robots being armed and autonomous. Using robots will reduce the risks so much that future wars will increase. Robots themselves have the inability to differentiate between combatants and noncombatants, which means there is a potential for higher civilian casualties.

2. *Decision Making*: A main concern is the decision-making aspect of robotic warfare. Should robots be able to make autonomous decisions about killing human beings? Or should humans continue to make the final decisions? Lawmakers, government officials, and society must decide how robots are employed in the Army. Military robots perform worse when humans would not stop interrupting them.

3. *Responsibility*: There is a possibility that machines may one day reach a point where they make more ethical decisions on the battlefield .Who is responsible when autonomous robots make mistakes? If their mistakes lead to fatalities, how do you hold robots accountable? Depending on policies, rule of law, and rules of engagement, it may be difficult to hold anyone accountable, especially if robots make independent decisions.

   The human creators and operators of autonomous robots must be held accountable for the machines' actions. They should exercise appropriate levels of judgment and remain responsible for the development, deployment, use, and outcomes of military systems.

4. *Security*: Concerns have been expressed on the security of autonomous systems compared to the remotely piloted systems currently in use. It is only a matter of time until we hand weapons over autonomous system that can outperform soldiers. The more the system is autonomous, the more it has the capacity to make choices other than those predicted or encouraged by its programmers. The intention of engineers in designing is that robots will make less mistakes than humans do in the battlefield. Robots need to be programmed with rules about when acceptable to fire on a tank and must learn how to perform complicated, emotionally fraught tasks, such as distinguishing civilians. Lately, robots have been outfitted with teleoperated weaponry, keeping them from killing individuals all alone.

5. *Fear*: Fear of the unknown and the humane side of war driven because the possibility of reducing risk to human troops may escalate new conflicts. What if a robot army becomes a weapon of mass destruction? The fear that hackers could take control of robots and AI, leading to mass death and destruction, is among the reasons lawmakers and government officials oppose such technology.

6. *Liability*: Robot autonomy is a critical ethical challenge. It is unclear who is responsible if an autonomous robot makes a mistake. Military robots are designed to follow the rules and conduct of the professions or roles they emulate, and it is expected that ethical principles are applied and aligned with such roles.

7. *Recruitment*: The use of robots could reduce the need for human soldiers, which could make it harder to recruit.

8. *Technological Advantage*: Nations with advanced robotic technology could have an unfair advantage in war.

3.9 CONCLUSION

Military robots are still in their early days. Today's modern military forces are using different kinds of robots for different applications ranging from mine detection, surveillance, logistics and rescue operations. Modern armed conflict can be fundamentally changed in the age of computers and networks. They can and will play a major role in the future of warfare; it is just a matter of when. The decision to use robots requires significant research, planning, and deliberate execution. The Pentagon has also started work on a number of unmanned systems for potential use in space, although most are still only on the drawing boards.

The use of robots in the military is only going to increase in the coming years. Robots are the future of the military. Robotics is the key to the future combat systems. For those looking to have a successful career in robotics, the military is a great option. They provide the training and on-the-job experience necessary to work with robots.

For more information about robotics in the military, one should consult the books in [15-34] and the following related journals devoted to robotics:

- *Robotica*
- *Robotics*
- *Robotics and Autonomous*
- *Robotics and Computer-Integrated Manufacturing*
- *Advanced Robotics*
- *Autonomous Robots*
- *Journal of Robotics*
- *Journal of Robotic Systems*
- *Journal of Robotic Surgery*
- *Journal of Robotics and Mechatronics*
- *Journal of Intelligent & Robotic Systems*
- *Journal of Mechanisms and Robotics-Transactions of the ASME*
- *Journal of Automation, Mobile Robotics and Intelligent Systems*
- *Journal of Future Robot Life*
- *IEEE Robotics and Automation Letters*
- *IEEE Transactions on Robotics*
- *International Journal of Medical Robotics and Computer Assisted Surgery*
- *International Journal of Robotics Research*
- *International Journal of Social Robotics*
- *International Journal of Humanoid Robotics*
- *International Journal of Advanced Robotic Systems*
- *Science Robotics*
- *Soft Robotics*
- *Military Review*
- *Journal of Military Learning*
- *NCO Journal*

REFERENCES

[1] S. Brezgov, "Robots in education: Is the educational revolution just around the corner?" January 2020, https://scholarlyoa.com/robots-in-education-is-the-educational-revolution-just-around-the-corner/

[2] M. N. O. Sadiku, U. C. Chukwu. A. Ajayi-Majebi, and S. M. Musa, "A primer on robotics, " *International Journal of Trend in Scientific Research and Development,* vol. 6, no. 7, November-December 2022, pp. 614-621.

[3] M. N. O. Sadiku, P. A. Adekunte, and J. O. Sadiku, "Robotics in the military," *International Journal of Trend in Scientific Research and Development*, vol. 8, no. 5, September-October 2024, pp. 170-179.

[4] "Human–robot interaction," *Wikipedia,* the free encyclopedia https://en.wikipedia.org/wiki/Human–robot interaction

[5] "Robotics," *Wikipedia*, the free encyclopedia https://en.wikipedia.org/wiki/Robotics

[6] "7 Types of robots used in the military," January 2023, https://www.adsinc.com/news/7-types-of-robots-used-in-the-military

[7] J. Detsch, "America's next soldiers will be machines," April 2024, https://foreignpolicy.com/2024/04/06/us-army-military-robots-soldiers-technology-testing-war/

[8] "The future of warfare: Robotic mules in military science fiction," https://milsf.com/the-future-of-warfare-robotic-mules-in-military-science-fiction/

[9] E. Ackerman and E. Guizzo, "How Stanford built a humanoid submarine robot to explore a 17th-century shipwreck," December 2016, https://spectrum.ieee.org/stanford-humanoid-submarine-robot

[10] L. Ross, "6 Ways the military is using robotics," March 2023, https://www.thomasnet.com/insights/6-ways-the-military-is-using-robotics/

[11] J. Judson, "The robots are coming: US Army experiments with human-machine warfare," March 2024, https://www.defensenews.com/unmanned/2024/03/25/the-robots-are-coming-us-army-experiments-with-human-machine-warfare/

[12] "Defence Rover to fight against terrorism and natural disasters," December 2021, https://www.electronicwings.com/users/ChiranjeeviGowda/projects/1672/defence-rover-to-fight-against-terrorism-and-natural-disasters

[13] "Revolutionizing defense: The role of robotics in military operations," February 2024, https://www.nextechsol.com/revolutionizing-defense-the-role-of-robotics-in-military-operations

[14] "The future is here: How the military uses robots," October 2019, https://community.robotshop.com/blog/show/the-future-is-here-how-the-military-uses-robots

[15] M. N. O. Sadiku, *Robotics and Its Applications.* Moldova, Europe: Lambert Academic Publishing, 2023.

[16] T. Cooke, *A Timeline of Military Robots and Drones.* Capstone, 2017.

[17] E. Noll, *Military Robots.* Bellwether Media, Incorporated, 2017.

[18] S. Price, *Amazing Military Robots.* Capstone, 2013.

[19] P. W. Singer, *Wired for War: The Robotics Revolution and Conflict in the 21st Century.* Penguin Books, 2009.

[20] A. Rossiter and P. Layton, *Warfare in the Robotics Age.* Lynne Rienner Publishers, 2024.

[21] R. Snedden, *Robotics in the Military.* Greenhaven Publishing LLC, 2017.

[22] D. R. Faust, *Military and Police Robots.* PowerKids Press, 2016.

[23] L. Spilsbury and R. Spilsbury, *Robots in the Military.* Gareth Stevens Publishing, 2015.

[24] F. Jentsch, *Human-Robot Interactions in Future Military Operations.* Boca Raton, FL: CRC Press, 2016.

[25] P. J. Springer, *Outsourcing War to Machines: The Military Robotics Revolution.* ABC-CLIO, 2018.

[26] P. J. Springer, *Military Robots and Drones: A Reference Handbook.* ABC-CLIO, 2013.

[27] V. Nath and S. E. Levinson, *Autonomous Military Robotics.* Springer, 2014.

[28] K. W. Larson, Military Robots. Creative Company, 2018.

[29] J. Galliott, *Military Robots: Mapping the Moral Landscape.* Taylor & Francis, 2016.

[30] B. S. Alpert, *Military Robots.* Capstone, 2019.

[31] L. Idzikowski, *Military Robots.* Lerner Publishing Group, 2023.

[32] F. Sabry, *Military Robot: Revolutionizing Warfare with Autonomous Technology.* One Billion Knowledgeable, 2024.

[33] A. Krishnan, *Killer Robots: Legality and Ethicality of Autonomous Weapons.* Taylor & Francis, 2016.

[34] S. D. White and S. White, *Military Robots.* Children's Press, 2006.

CHAPTER 4

DRONES IN THE MILITARY

"In the age of the Almighty Computer, drones are the perfect warriors. They kill without remorse, obey without kidding around, and they never reveal the names of their masters."
– Eduardo Galiano

4.1 INTRODUCTION

The United States has been at the forefront of military drone development and deployment. The number of nations using drones has increased to about 50 in recent years, including China and Iran. The primary purpose of the Department of Defense (DoD) domestic aviation operations are to support Homeland Defense (HD) and Defense Support of Civilian Authorities (DSCA) operations, and military training and exercises. The primary purpose of DoD domestic UAS operations is for DoD forces to gain realistic training experience, test equipment, and tactics in preparation for potential overseas warfighting missions. The vast majority of DoD UAS training is conducted in airspace delegated by the FAA for DoD use [1]. While drones have their many civilian uses in agriculture, education, business, manufacturing, surveillance, film making, etc. military drones are armed ones used in combat.

Drones have rapidly evolved into an essential component of modern warfare. Today, drones primarily serve the military industry around the world. Military drones have become indispensable tools for special operations forces, providing real-time situational awareness, communication relay, and electronic warfare capabilities. They can be used to establish communication networks in the battlefield, relaying signals between ground forces and command centers. Military drone applications have revolutionized the way modern warfare is conducted. From surveillance to combat and logistics, drones have become an indispensable tool for military operations worldwide [2].

This chapter examines the applications of drones in the military. It begins with explaining what a drone is. It discusses military drones. It provides some applications of military drones. It highlights the benefits and challenges of military drones. It concludes with comments.

4.2 WHAT IS A DRONE?

The FAA defines drones, also known as unmanned aerial vehicles (UAVs), as any aircraft system without a flight crew onboard. Drones include flying, floating, and other devices, including unmanned aerial vehicles (UAVs), that can fly independently along set routes using an onboard computer or follow commands transmitted remotely by a pilot on the ground. A typical drone is shown in Figure 4.1 [3]. A drone is usually controlled remotely by a human pilot on the ground, as typically shown in Figure 4.2 [4]. Drones can range in size from large military drones to smaller drones. Drones, previously used for military purposes, have started to be used for civilian purposes since the 2000s. Since then, drones have continued to be used in intelligence, aerial surveillance, search and rescue, reconnaissance, and offensive missions as part of the military Internet of things (IoT). Today, drones are used for different purposes such as aerial photography, surveillance, agriculture, entertainment, healthcare, transportation, law enforcement, etc.

Figure 4.1 A typical drone [3].

Figure 4.2 A drone is usually controlled by operators on the ground [4].

Commercial drones have come a long way in the last decade. Drones work much like other modes of air transportation, such as helicopters and airplanes. When the engine is turned on, it starts up, and the propellers rotate to enable flight. The motors spin the propellers and the propellers push against the air molecules downward, which pulls the drone upwards. Once the drone is flying, it is able to move forward, back, left, and right by spinning each of the propellers at a different speed. Then, the pilot uses the remote control to direct its flight from the ground [5].

Drone laws exist to ensure a high level of safety in the skies, especially near sensitive areas like airports. They also aim to address privacy concerns that arise when camera drones fly in residential areas. These include the requirement to keep your drone within sight at all times when airborne. In the United States, drones weighing less than 250g are exempt from registration with civil aviation authorities. If your drone exceeds 250g in weight, you will also require a Flyer ID, which requires passing a test [6]. It is necessary to register as an operator, be trained as a pilot, and have civil liability insurance, in addition to complying with various flight regulations, and those of the places where their use is permitted.

Most drones have a limited payload, usually under 11 pounds. Drones are classified according to their size. Here are the different drone types:

- Nano Drone: 80-100 mm
- Micro Drone: 100-150 mm
- Small Drone: 150-250 mm
- Medium Drone: 250-400 mm
- Large Drone: 400+ mm

One of the emerging trends in drone use for factories is the utilization of LiDAR technology. LiDAR stands for Light Detection and Ranging. This technology provides accurate depth information essential for understanding the three-dimensional structure of the environment. LiDAR sensors emit laser beams to measure distances to objects, creating high-resolution 3D maps of the surrounding terrain and objects. The ability to capture detailed data through LiDAR technology has opened up opportunities for better predictive maintenance, reduction in inspection times, and overall cost savings [7].

4.3 MILITARY DRONES

The concept of unmanned aerial vehicles(UAVs) dates back to the early 20th century. But it was not until World War II that UAVs began to take shape as a vital military tool. In the late 20th century and early 21st century, military drones evolved rapidly. Today, military drones are used by numerous nations for intelligence, surveillance, and reconnaissance (ISR) missions. In 1959, as tensions between the US and the Soviet Union began to skyrocket, so did drone innovation. The first military conflict in which UAVs played a major role was the Gulf War, after which military drones became commonplace worldwide.

Military drones are remote-controlled aircraft of various sizes designed to perform tasks deemed too dull, dirty, or dangerous for human troops. The main selling point of a military drone is the lack of an on board human pilot. The drones have revolutionized the way modern warfare is conducted by providing a means of engaging targets with precision and reduced risk to military personnel. They are often used in dangerous areas where human soldiers might be at risk. These drones can conduct precision strikes on high-value targets. A typical military drone is shown in Figure 4.3 [8]. Military drones are remotely-piloted UAVs used for monitoring, mapping, target acquisition, intelligence, battle damage management, and surveillance. They are armed with missiles, bombs, or anti-tank weapons. These drones have been a valuable asset in the military for many years. Pentagon takes the issue of military drones used on American soil very seriously.

Figure 4.3 A typical military drone [8].

Figure 4.4 shows different types of military drones [9]. The types include the following [10]:

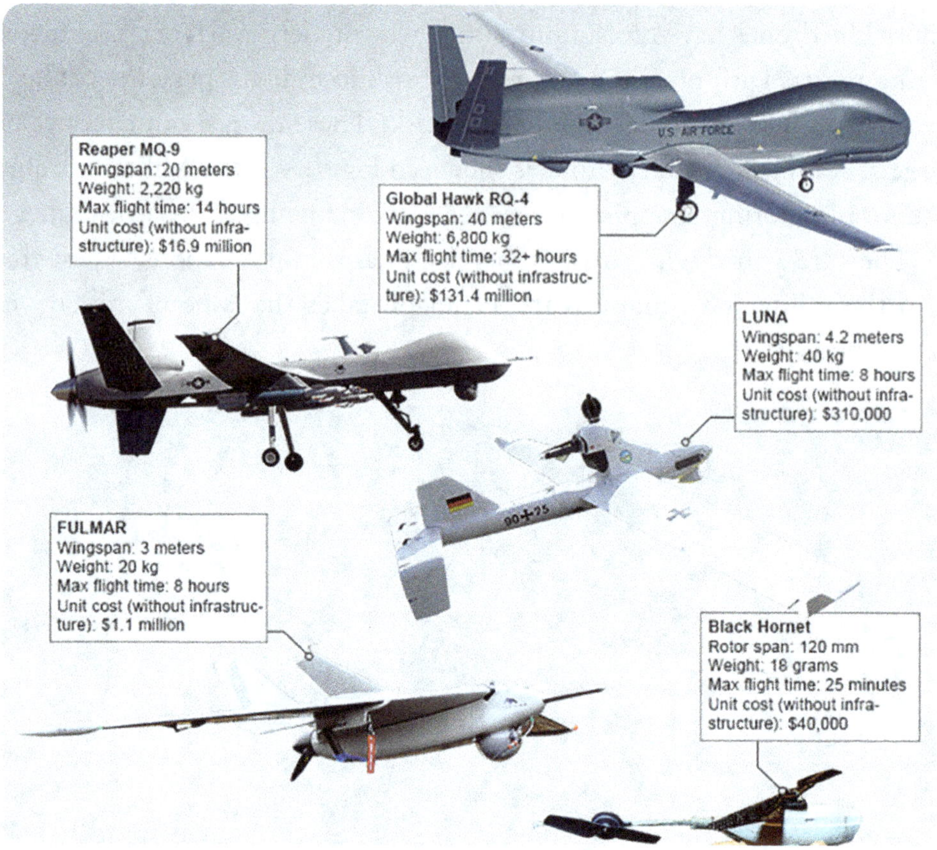

Figure 4.4 Different types of military drones [9].

- Combat drones (or fighter drones) are designed for offensive operations.
- Tactical drones are used in a range of specialized missions, such as communication relay, electronic warfare, and counter-drone operations. Figure 4.5 shows an example of tactical drone [11].

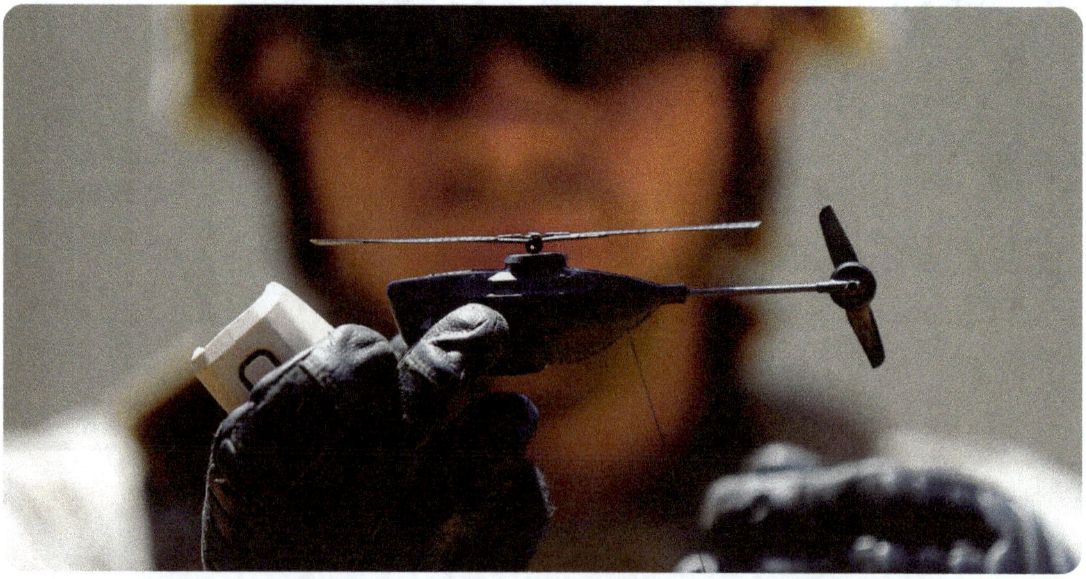

Figure 4.5 An example of tactical drone [11].

- Transport drones are designed for logistical support, such as delivering supplies, medical evacuation, and personnel transport.
- Reconnaissance drones can be used to monitor enemy movements, troop deployments, and infrastructure.
- Surveillance drones have the ability to observe enemy activities without putting human lives at risk.
- Detecting and tracking drones can provide real-time information on drone locations, flight paths, and potential payloads.
- Swarming drones refer to coordinated deployment of multiple drones to carry out complex tasks or missions.
- Logistics drones are drones that are used for logistical purposes, such as transporting supplies and equipment to troops in the field.
- Target drones are drones that are used as targets for training exercises.
- Stealth drones are drones that are designed to be stealthy and difficult to detect.

Leading manufacturers in the global market of military drones include [12]:

1. General Atomics Aeronautical Systems, Inc.
2. Northrop Grumman Corporation
3. Israel Aerospace Industries Ltd
4. BAE Systems Plc
5. Lockheed Martin Corporation
6. Raytheon Company
7. Insitu Inc.
8. AeroVironment, Inc.
9. Turkish Aerospace Industries, Inc.
10. Elbit System Ltd
11. The Boeing Company
12. Thales Group
13. Saab Group

Figure 4.6 shows 3D-printed military drones [13].

Figure 4.6 3D-printed military drones [13].

4.4 APPLICATIONS

Military drone applications have expanded dramatically, playing a critical role in intelligence gathering, combat operations, and logistical support. Drones are used by the military for a variety of purposes, including surveillance, target acquisition, reconnaissance, logistics, and targeted strikes. Common applications of drones in the military include the following:

- *Electronic Warfare:* Electronic warfare is the art of locating enemy forces by the signals that they send out and then isolating them by jamming their communications. This is a critical aspect of modern military operations, and tactical drones have emerged as valuable assets in this domain. They can be equipped with electronic warfare systems to detect, identify, and disrupt enemy radar and communication systems, rendering them ineffective. Drones can jam enemy communications and detect radar installations.

- *Enforcement and Countermeasures:* Law enforcement agencies are also investing in counter-drone technologies and capabilities to detect and neutralize unauthorized or malicious drone activities. These efforts include jamming and spoofing. Once a drone has been detected, jamming and spoofing techniques can be used to disrupt its communication and control systems. Jamming involves the transmission of radio frequency signals to interfere with the drone's remote control or GPS navigation, while spoofing involves sending false signals to deceive the drone's GPS receiver, causing it to deviate from its intended course or crash.

- *Surveillance:* Drones' surveillance capabilities allow for better situational awareness on the battlefield, helping to detect potential threats and plan effective military strategies. In urban warfare scenarios, where the environment is complex and unpredictable, drones can be invaluable assets, assisting ground troops in navigating and understanding the terrain from above.

- *Reconnaissance:* Drones can conduct surveillance missions by hovering over an area for an extended period. They can survey hostile territories and provide real-time information to command centers. This can help protect soldiers from danger and improve situational awareness. They can provide rapid assistance that significantly impacts the survival and recovery of injured individuals. They can help swiftly locate and assess injured soldiers in combat zones.

- *Logistics*: Drones are revolutionizing the delivery of medical supplies. They can carry essential medical equipment and medications to injured personnel in areas inaccessible by traditional means. They can deliver supplies to front-line units in hard-to-reach areas. They can also help evacuate injured personnel. A drone can hover around and land anywhere without using runways. Examples are retail companies, parcel services, and pharmacies using drone delivery to bring goods and medications to our doorstep. Manufacturers are also using drones to ship supplies directly to businesses. Figure 4.7 shows a delivery drone [14].

Figure 4.7 A delivery drone [14].

- *Communication:* Drones can act as airborne communication relays to connect units and command centers, especially in areas with compromised communication infrastructure. Drones also use radio to send data like videos and receive remote control.

- *Search and Rescue:* Drones can be used for combat search and rescue. For example, imagine a search and rescue mission in a dense forest, where a drone uses its AI edge computing capabilities to navigate through the complex environment. Some drones for rescue missions are portrayed in Figure 4.8 [15].

Figure 4.8 Drones for rescue missions [15].

- *Autonomous Navigation:* Modern drones are equipped with cutting-edge AI and edge computing that enables them to navigate complex environments autonomously. This includes obstacle avoidance, terrain analysis, and even adaptive mission planning based on real-time data. It can identify and avoid obstacles like trees and cliffs, analyze the terrain to find the safest and most efficient routes, and adapt its flight plan in real-time based on changes in the environment.

- *Decision Making:* AI algorithms allow drones to make critical decisions rapidly. For instance, a drone can analyze threats, select targets, and even choose flight paths with minimal human input. In a military operation, a drone equipped with AI algorithms can rapidly analyze aerial footage to identify potential threats, such as enemy combatants or unsecured territories. It can then autonomously select the safest flight path to avoid detection or confrontation.

- *Coordinated Operations:* Swarm technology involves multiple drones working together, coordinated through advanced AI. This approach allows for complex, large-scale operations where drones collaborate to achieve common objectives. Swarm drones can be used in various scenarios, from reconnaissance missions to creating real-time 3D maps of battlefields to overwhelming enemy defenses. Figure 4.9 displays a swarm of drones [3].

Figure 4.9 A swarm of drones [3].

- *Military Exercises:* Drones are becoming increasingly important in modern military exercises. Military exercises are an important part of military training around the world. It is a way for soldiers to hone their skills, test their equipment and improve their tactics. In recent years, drone technology has played a key role in military exercises. Drones are a game-changer when it comes to military operations and can help soldiers in numerous ways. Drones are also being used in more specialized training exercises. They can be used to train in complex and dangerous environments that would be too dangerous or expensive to train in with live troops. The use of drones in military exercises is not without its challenges. One challenge is that drones can be easily hacked or jammed.

4.5 BENEFITS

Drones play a crucial role in war, medicine, and rescue missions. Their success in different operations makes them an invaluable asset. Drone strikes ensure the safety of the United States by deconstructing terrorist organizations anywhere in the world. They protect more US military personnel and kill fewer civilians than any other weapon used by the military. They are cheaper than manned aerial warfare or ground combat. The Pentagon has deployed drones to spy over US territory for non-military missions over the past decade. Other benefits of military drones include the following [16]:

- *Better Reconnaissance:* Drones provide real-time information on targets' positions, terrain, and enemy movements to commanders on the ground. Compared to high-altitude aircraft, drones can take closer footage without compromising the quality of both photos and video.

- *Reduced Cost:* Drones are cheaper than conventional aircraft in terms of both price and maintenance. Because drones are unmanned, they also reduce the risk of pilots being injured mid-flight. States and non-state groups that cannot afford to buy fighter jets can buy drones.

- *Increased Convenience:* Compared to conventional aircraft, drones are faster and easier to deploy. They are easier to operate and do not need training as extensive as most aircraft. Also, many drones do not need a runway, and other types can easily fit in a backpack.

- *Enhanced Safety:* Military drones have changed how military bases secure their perimeters and monitor for threats, improving surveillance and soldier safety. Drone operators can provide real-time information without putting themselves at risk. On top of this, that same information also informs commanders where to position their troops to ensure safety. Everything the FAA does is focused on ensuring the safety of the nation's aviation system.

- *Increased Flexibility:* Military forces always need to be ready for anything at a moment's notice. Drones are helpful in readiness. They can even be fully automated. They provide many benefits and advantages that make them extremely useful for different roles.

- *Precision Warfare:* Armed drones have become a game-changer in modern warfare. With the ability to carry various types of munitions, including missiles and guided bombs, they offer precise and lethal firepower. Precision warfare, made possible by drone technology, minimizes collateral damage and civilian casualties, making it a more ethically acceptable form of warfare compared to traditional indiscriminate bombardments.

4.6 CHALLENGES

Critics of the drone strikes argue that drone strikes wreak havoc on civilian communities and in turn create more terrorists than they set out to destroy. Drone operations are secretive, lack sufficient legal oversight, and prevent citizens from holding their leaders accountable. In spite of the growing use of drones, they remain a controversial and unpopular tactic. Other challenges of military drones include the following [16]:

- *Cost:* The Pentagon abhors cheapness; no production line exists for cheap drones or cheap artillery shells. The defense industry prices are prohibitively high. Official procurement figures are classified, but press reports indicate per unit costs for military drones vary from $6,000 to $58,000–– twelve to one hundred times more expensive than Ukraine's home-assembled drones. (Ukraine is producing one million drones). The same cost disparity affects defense just as much as offense on land. Another challenge is that drones can be expensive to operate.

- *Drone Strikes:* Drone strikes violate the sovereignty of other countries and are extremely unpopular in the affected countries. They often raise complex legal and moral dilemmas. They are illegal under both international and United States humanitarian law, which states that lethal force is only permissible when the target poses an immediate threat to the country's survival. And it can be argued that not all drone targets fit this category. The death toll from American drone strikes was approximately 2,400 in total from 2009-2014 and has risen to more than 6,000 since 2015.

- *War:* Whichever side you are fighting on, war is horrific and lamentable. Less of them are in harm's way because of the use of drones. The proliferation of drone technology poses security risks, as non-state actors and adversarial nations can utilize drones for malicious purposes. Drones have been linked with civilian deaths in many conflict zones

- *Ethical Concern:* The use of military drones, particularly in armed strikes, has raised several legal and ethical concerns. The concerns refer to the use of combat drones in targeted killings, that is the intentional killing of specific individuals outside of an active battlefield, resulting in the unintentional deaths of innocent civilians.

- *Accountability*: Another significant concern is the lack of transparency and accountability surrounding drone operations. Due to the covert nature of drone warfare, it can be challenging to ascertain the precise circumstances and justifications for drone strikes. This lack of accountability has led to calls for greater oversight of drone operations to ensure that they are conducted in accordance with international law and ethical standards.

- *Regulations:* To prevent the misuse of drones and ensure public safety, governments worldwide have implemented legal restrictions and regulations governing the use of civilian drones. These regulations typically include requirements for registration, flight restrictions in specific areas (e.g. near airports, military installations, or populated areas), and limitations on drone size, weight, and capabilities.

4.7 CONCLUSION

The rise of drones has been nothing short of revolutionary in the military industry. Modern warfare and national defense using drones make for big business the world over.

With the largest military budget in the world, the United States is a leader in the development and production of the drones.

As technology continues to evolve, so too will military drone applications. As the use of military drones becomes more prevalent, the development of counter-drone measures and defense systems has become increasingly important. An increase in terror threats, unconventional military threats, and geopolitical tensions worldwide have led to an increase in demand for unmanned aerial vehicles (UAVs) to target terrorist and insurgent groups across the globe. As the market for military drones continues to expand, driven by increasing government funding and technological innovations, drones are set to play an even more crucial role in shaping the landscape of global security and defense strategies.

The search for small disposable and accessible drones in military and defense is on the rise around the world. Every military specialist agrees that Unmanned Aerial Vehicles, or drones, are the future of warfare. Future drones will be even more autonomous, capable of performing complex tasks such as tactical strikes, surveillance missions, and supply deliveries without human intervention. More information about drones in the military can be found in the books in [17-34] and a related journal: *Drones*.

REFERENCES

[1] "Unmanned aircraft systems (UAS),"
https://dod.defense.gov/UAS/

[2] M. N. O. Sadiku, P. A. Adekunte, and J. O. Sadiku, "Drones in the military," *International Journal of Trend in Scientific Research and Development*, vol. 8, no. 5, September-October 2024, pp. 58-65.

[3] "The future of warfare and security: Drones in military and HLS applications," January 2024,
https://www.maris-tech.com/blog/the-future-of-warfare-and-security-drones-in-military-and-hls-applications/

[4] "The impact of drones on future of military warfare,"
https://media.inti.asia/read/the-impact-of-drones-on-future-of-military-warfare#:~:text= Here%20 are%20some%20additional%20thoughts,each%20other%20in%20real%20time.

[5] "How drones work and how to fly them," May 2024,
https://dronelaunchacademy.com/resources/how-do-drones-work/

[6] "What are the main applications of drones?" June 2024,
https://www.jouav.com/blog/applications-of-drones.html

[7] "Drones in manufacturing: A game-changer for industry,"
https://viper-drones.com/industries/infrastructure-drone-use/manufacturing/#:~:text=The% 20integration%20of%20drones%20into,on%20manufacturing%20is%20no%20exception.

[8] "Military drones UAS & UAVS for ISTAR,"
https://heighttechnologies.com/military-istar-drones/

[9] B. Knight, "A guide to military drones," June 2017,
https://www.dw.com/en/a-guide-to-military-drones/a-39441185

[10] C. Guarnera, "Overview of military drone applications," April 2023,
https://www.bluefalconaerial.com/overview-of-military-drone-applications/

[11] "Decoding military surveillance drones: the ultimate guide,"
https://elistair.com/military-surveillance-drones/

[12] "Top 13 military drone manufacturers in the world," January 2022, https://roboticsbiz.com/top-13-military-drone-manufacturers-in-the-world/

[13] "3D-printed military drones 'assembled in a matter of hours'," February 2023, https://www.imeche.org/news/news-article/3d-printed-military-drones-assembled-in-a-matter-of-hours

[14] "Advanced air mobility," https://www.gore.com/products/industries/aerospace/advanced-air-mobility?xcmp=aam aero ppc defense google na%7c%7cgeneral&s kwcid=AL!13180!3!647674247881!p!!g!!advanced%20air%20 mobility&gad source=1&gclid=EAIaIQobChMIh8O x efiAMVdm0PAh30cSprEAAYAiAAEg LTTfD BwE

[15] A. Maltsev, "Drones at war and computer vision," December 2023, https://medium.com/@zlodeibaal/drones-at-war-and-computer-vision-a16b8063be7b

[16] M. Abunuwara, "Military drones," https://militarymortgagecenter.com/us-military/aircraft/military-drones/

[17] H. Marcovitz, *Military Drones (World of Drones).* Referencepoint Press, 2020.

[18] P. J. Springer, *Military Robots and Drones: A Reference Handbook.* Bloomsbury Publishing, 2013.

[19] T. Cooke, *A Timeline of Military Robots and Drones.* Capstone, 2017.

[20] M. Scheff, *Military Drones.* Raintree, 2019.

[21] J. E. Jackson, *One Nation Under Drones: Legality, Morality, and Utility of Unmanned Combat Systems.* Naval Institute Press, 2018.

[22] C. Enemark, *Armed Drones and the Ethics of War: Military Virtue in a Post-heroic Age.* Taylor & Francis, 2013.

[23] A. Završnik, *Drones and Unmanned Aerial Systems: Legal and Social Implications for Security and Surveillance.* Springer, 2015.

[24] D. Cortright, K. Wall, and R. Fairhurst (eds.), *Drones and the Future of Armed Conflict: Ethical, Legal, and Strategic Implications.* University of Chicago Press, 2017.

[25] J. Kaag and S. Kreps, *Drone Warfare.* Polity Press, 2014.

[26] A. Stilwell, *Military Drones: Unmanned Aerial Vehicles (UAV).* Amber Books Limited, 2023.

[27] S. J. Frantzman, *The Drone Wars: Pioneers, Killing Machines, Artificial Intelligence, and the Battle for the Future.* Bombardier Books, 2021.

[28] D. Sloggett, *Drone Warfare: The Development of Unmanned Aerial Conflict.* Skyhorse Publishing, 2015.

[29] E. D. Nucci and F. S. de Sio, *Drones and Responsibility: Legal, Philosophical and Socio-Technical Perspectives on Remotely Controlled Weapons.* Taylor & Francis, 2016.

[30] M. Schuh, *Military Drones and Robots.* Capstone, 2022.

[31] D. R. Faust, *Military Drones.* Rosen Publishing Group, 2015.

[32] C. P. McCarthy, *Military Drones.* ABDO Publishing Company, 2020.

[33] J. I. Walsh and M. Schulzke, *Drones and Support for the Use of Force.* University of Michigan Press, 2018.

[34] M. Chandler, *Military Drones.* Capstone, 2019.

CHAPTER 5

3D PRINTING IN MILITARY

"3D printing represents the democratization of manufacturing.
It allows anyone to create anything, anywhere, at any time."
– Neil Harbisson

5.1 INTRODUCTION

The field of defense technology is constantly evolving, and one groundbreaking innovation that has emerged in recent years is 3D printing. This technology has the potential to transform the way missiles are produced. The promise of cost reductions in the production of tools and components for military equipment, to design flexibility and on-demand, on-site manufacturing, 3D printing allows military pundits a lot of liberty. Figure 5.1 shows how the world's largest concrete 3D printer constructs barracks for the US Army in Champaign, Illinois [1].

Figure 5.1 World's largest concrete 3D printer constructs barracks [1].

Although the USA pioneered the adoption of 3D printing in defense, the list of countries incorporating this technology continues to grow. Several nations, such as the USA, UK, Spain, Canada, Australia, and France have significantly invested in the use of additive manufacturing for the development of military equipment, vehicles, and weapons. They are pouring billions into 3D printing and related technologies.

3D printing, also known as additive manufacturing, is a process of creating three-dimensional objects by adding material layer by layer based on a digital model. It has come a long way since it was first developed in the 1980s. As in other industries, 3D printing is making its case within defense and military fields. The technology is revolutionizing the way militaries around the world procure and maintain their equipment. As 3D printing took off in recent years in the defense sectors additive manufacturing companies have been working to drive economies of scale. The military can benefit from 3D printing since the military personnel are exposed to all sorts of environments around the globe [2].

This chapter takes a look at how the defense industry has embraced 3D printing. It begins with explaining what 3D printing is all about. It presents 3D printing in the defense industry. It provides several applications of 3D printing in defense. It highlights the benefits and challenges of employing 3D printing in defense industry. It concludes with some comments.

5.2 WHAT IS 3D PRINTING?

3D printing (also known as additive manufacturing (AM) or rapid prototyping (RP)) was invented in the early 1980s by Charles Hull, who is regarded as the father of 3D printing. Since then it has been used in manufacturing, automotive, electronics, aviation, aerospace, aeronautics, engineering, architecture, pharmaceutics, consumer products, education, entertainment, medicine, space missions, the military, chemical industry, maritime industry, printing industry, and jewelry industry [3]

A 3D printer works by "printing" objects. Instead of using ink, it uses more substantive materials–plastics, metal, rubber, and the like. It scans an object–or takes an existing scan of an object–and slices it into layers, which can then convert into a physical object. Layer by layer, the 3D printer can replicate images created in CAD programs. In other words, 3D printing instructs a computer to apply layer upon layer of a specific material (such as plastic or metal) until the final product is built. This is distinct from conventional manufacturing methods, which often rely on removal (by cutting, drilling, chopping, grinding, forging, etc.) instead of addition. Models can be multi-colored to highlight important features, such as tumors, cavities, and vascular tracks. 3DP technology can build a 3D object in almost any shape imaginable as defined in a computer-aided design (CAD) file. It is additive technology as distinct from traditional manufacturing techniques, which are subtractive processes in which material is removed by cutting or drilling [4].

3D printing has started breaking through into the mainstream in recent years, with some models becoming affordable enough for home use. Many industries and professions around the world now use 3D printing. It plays a key role in making companies more competitive. The gap between industry and graduating students can be bridged by including the same cutting-edge tools, such as 3D printing, professionals use every day into the curriculum. There are 3D printed homes, prosthetics, surgical devices, drones, hearing aids, and electric engine components. As shown in Figure 5.2, 3D printing involves three steps [5]. A typical 3D printer is shown in Figure 5.3 [6].

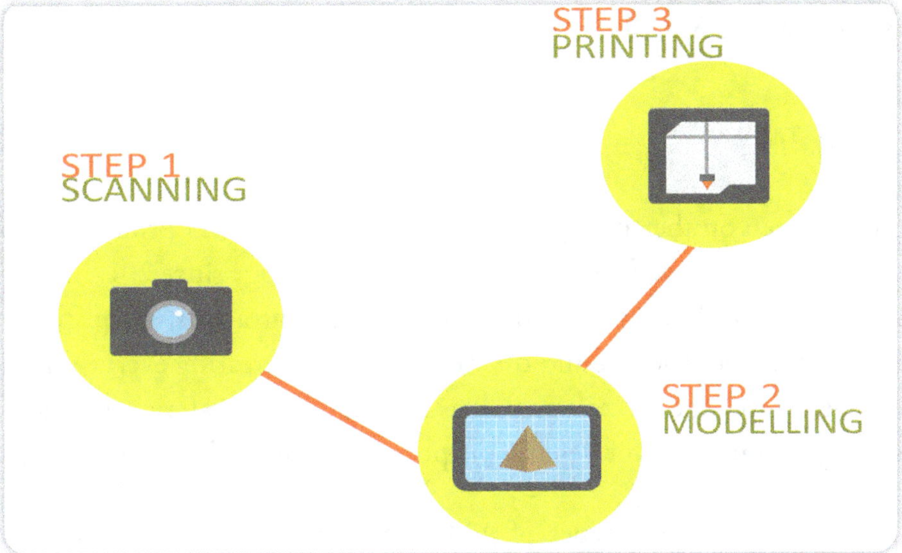

Figure 5.2 3D printing involves three steps [5].

Figure 5.3 A typical 3D printer [6].

The use of additive manufacturing has expanded through medical devices, aerospace, automotive engineering, fashion, construction, and other fields. Numerous 3D printing companies offer ready-made menus of different materials and techniques. Several aerospace and defense companies have already adopted this technology in their operations. The advancement of 3D printing within aerospace and defense is in large part driven by key industry players, including GE, Airbus, Boeing, Safran, GKN, Stratasys, Latécoère, and Moog Aircraft Group.

5.3 3D PRINTING IN MILITARY

The military has always been on the cutting edge of technology, and 3D printing is no exception. The US military has been involved with additive manufacturing technology since 2012 when 3D printers were first deployed in the field by the Army, Navy, and various DoD contractors. 3D printing is being used by military units from different countries around the world. They are creating everything from replacement parts for critical vehicles, ships, and aircraft as well as new designs for safety equipment. It is critical for them to develop the right equipment for defense purposes, a matter of life and death in the battlefield.

While there are now several different types of AM, each forms an object in the same basic way: A programmed machine adds material, layer by layer, until a three-dimensional object is formed. Few technologies are as vigorously hyped throughout the military as additive manufacturing, for good reason: The ability to produce components on demand, at the point of need, will transform logistics, reduce material waste, and enable customization, all at a fraction of the costs and times involved in the manufacturing process than traditionally techniques. New designs can be prototyped and tested rapidly, without having to stand up production lines or create expensive tooling.

Contrary to traditional manufacturing methods, 3D printing offers flexibility and responsiveness, particularly in addressing the needs of the aerospace and defense sectors. The beauty of additive engineering, design, and manufacturing products is the fact that you have that vital thread from when you first start designing the part, analyzing it, doing topology optimization, light-weighting, printing, finishing and inspecting. While much of the world continues to think of AM technology as a convenient way to make sturdy plastic objects from 3D printers, military personnel at all levels have been pushing its limits far beyond what most imagined possible. The selection of available 3D printable materials for aerospace and defense applications ranges from engineering-grade thermoplastics to metal powders. Speed of use and versatility sum up what makes additive manufacturing different. When there is a need to build something state-of-the-art we do that and are comfortable with the risk. Manufacturing parts through a 3D printer can cut down on time and cost in comparison to ordering specialized parts.

5.4 APPLICATIONS OF 3D PRINTING IN MILITARY

There are countless ways in which the military is using 3D printing to improve the workflow of their service members. Defense professionals have also been using 3D printing to create bunkers, vehicle hide structures, bridges and buildings in various locations around the world to serve to support and house military personnel and machinery at bases as well as in the field. All branches of the US military (the US Army, Air Force, Navy, and Marines) use 3D printing. They are exploring more efficient and effective ways to 3D print anywhere in the field whether on board a ship at sea or at a remote base camp. Some of the common applications of 3D printing in defense industry include the following [7,8]:

1. *Prototyping:* The ability to 3D print a prototype and make adjustments in a matter of minutes is crucial. A prototype can be printed multiple times to the same specifications. Military prototypes can be 3D printed quickly and cheaply. 3D printing technology is being used by the military to create prototypes and design new equipment. Additive manufacturing is an ideal solution for creating quick concept models and prototypes. It is widely used in the defense industry to rapidly produce prototypes without the need for expensive tooling. Design concepts, as well as validation testing, can be done much faster, thereby shortening the product development cycle. For example, 3D printing has made manufacturing weapons become easier, as shown in Figure 5.4 [9].

Figure 5.4 3D printing has made manufacturing weapons become easier [9].

2. *Replacement Parts:* Perhaps the most crucial advantage of choosing 3D printing technology over traditional manufacturing techniques lies in spare parts production, enabling highly efficient maintenance and repair of military defense systems. Since the average lifespan of an aircraft can range between 20 and 30 years, making maintenance and repair become important functions in the industry. Now, cut down production time from weeks, and even months, to a matter of hours. The defense industry relies heavily on spare and replacement parts. 3D printing becomes a viable option, as it facilitates the cost-effective and relatively quick production of parts and tools on demand, helping to maintain an aging fleet without having to order a spare part or to return to a port. As replacement parts can quickly run out of stock, 3D scanning techniques can be used to reverse engineer an available part, which then can be easily replicated via 3D printing. Figure 5.5 shows 3D-printed spare parts [10], while Figure 5.6 portrays solder making repair incorporating 3D-printer parts [11].

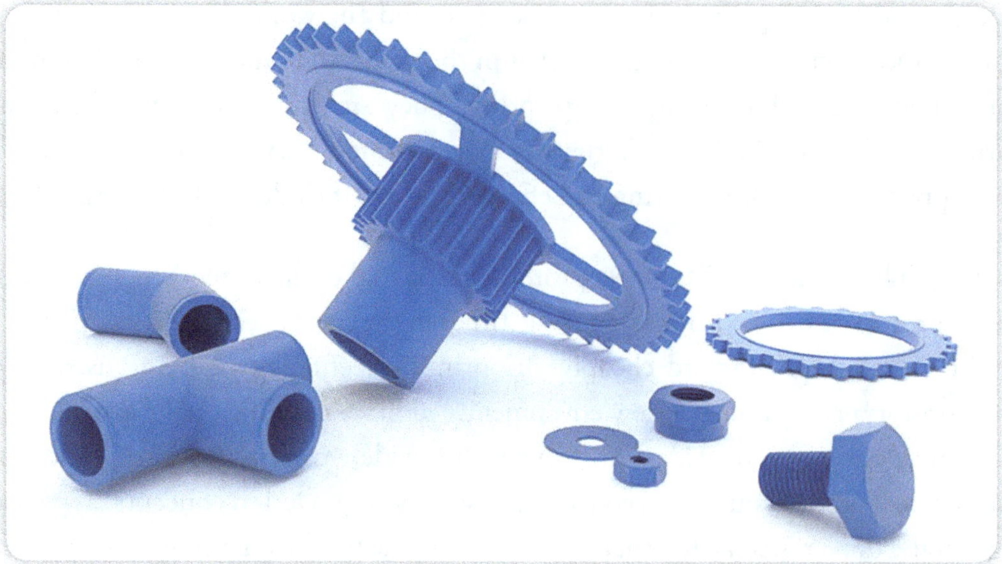

Figure 5.5 Army 3D-printed spare parts [10].

Figure 5.6 Solders making repair incorporating 3D-printer parts [11].

3. *Structural Components:* 3D printing can make a significant impact in the production of end parts for military equipment. The applications of additive manufacturing here vary, from complex brackets and small surveillance drones, to jet engine components and submarine hulls. The US Army has improved the impact absorption of the Army's combat helmet through 3D printing by incorporating advanced lattice geometries. Figure 5.7 shows a 3D-printed structural component [12].

Figure 5.7 A 3D-printed structural component [12].

4. *Metal 3D Printing*: Companies have used 3D print technology to print defense and military equipment in a variety of metals, irrespective of the complexity of the design. This is a huge shift from conventional manufacturing which rendered many complex designs useless due to its inability to produce it. Now, 3D print military weapons such as guns, rocket launchers, drones, and even missiles can be produced using 3D printing and materials such as steel, stainless steel, aluminum, titanium, nickel alloy, and cobalt chrome. Figure 5.8 shows solders using 3D printing [13].

Figure 5.8 Solders are using 3D printing [13].

5. *Plastic 3D Printing:* Plastic 3D printing has proven itself to be critical for the purposes of prototyping, research and development, and restocking in defense. Military operations will be handicapped without supplies such as food, fuel for vehicles, ammunition and tools for repair and maintenance. Materials used include nylon, PC, and Alumide.

6. *Medical Supply:* Another area where the military is using 3D printing is to create medical supplies. This is especially important in combat situations, where time is of the essence. 3D-printed medical supplies can be produced quickly and easily in the field, which means soldiers can be treated faster and have a higher chance of survival in combat situations. Additive manufacturing can also play a key role in the military medical sector by providing customized implants, prosthetics, and medical tooling in the field.

7. *3D Printed Structures:* Defense professionals have also been using 3D printing to create bunkers, vehicle hide structures, bridges and buildings in various locations around the world to serve to support and house military personnel and machinery at bases as well as in the field. The military has partnered with civilian companies to start building 3D-printed barracks to house service members for training missions. They have also been using giant robotic 3D printing to create bunkers, vehicle hide structures, bridges and buildings in various locations around the world to serve to support and house military personnel and machinery at bases as well as in the field. Figure 5.9 is an example of 3D-printed building [14].

Figure 5.9 A 3D-printed building [14].

8. *Armor Manufacturing:* As 3D printing technology continues to advance, new frontiers are being explored in the field of armor and military applications. The technology can be used to create mass-customized equipment for the military, such as armor suits that fit comfortably on the wearer. It allows for the creation of specialized tools for landmine clearance and explosive disposal.

5.5 BENEFITS

3D printing has the potential to transform the defense industry, providing new ways to 3D print replacement parts on demand, while reducing production costs for tools and components, and enabling new design engineering possibilities. Other benefits of 3D printing in defense include the following [8]:

1. *Faster Product Development:* 3D printing significantly speeds up the design process, since it requires no tooling. It cuts down production time from weeks, and even months, to a matter of hours. In contrast, traditional manufacturing can take months to produce the necessary tools to create end parts and prototypes. The defense industry can capitalize the technology to bypass costly and time-consuming tooling, thereby reducing the time required for product development.

2. *Freedom of Design:* Designers now have the freedom to cut down on unnecessary features on equipment that drastically helps to reduce overall product weight. The defense industry can take advantage of 3D printing's ability to produce freeform, optimized objects. Leveraging advanced design tools, design engineers can reduce the number of components in an assembly to just one, and thus greatly simplify the assembly process.

3. *Customization:* 3D printing offers the opportunity to create customized parts, tailored for specific functions. Soldiers can use 3D printing systems to manufacture customized parts on demand. Customizable 3D printable designs are vital to achieving greater levels of agility and flexibility within the military.

4. *On-Demand Production:* A major advantage of 3D printing is that it allows for on-demand manufacturing, which means that militaries can print the items they need when they need them, especially on the battlefield where time is often of the essence and sometimes lives are at stake. With 3D printing, the military can create the parts they need on-site whether that is on base, at the front lines, or at sea. Coordinating logistics and transportation makes up a significant part of any military budget. It is more cost-effective to print custom parts, tools, and spare parts near the point of use, a solution which additive manufacturing provides. This implies that soldiers in remote areas can also use 3D printing to their advantage.

5. *Waste Reduction:* 3D printing reduces material cost for the defense industry and also reduces the waste produce, thereby increasing material efficiency. As the making of defense products requires costly raw materials such as titanium, so when these raw materials are wasted the amount of money that goes in vain hampers budget of the company. Traditional methods of manufacturing defense products are not at affordable for majority of the companies but additive manufacturing definitely saves quite some amount of expenses.

6. *Cost Reduction:* 3D printing eliminates the cost of tooling and setup. This makes the entire process way more economic and feasible. Not only does additive manufacturing makes the production

cheaper, but it also saves ample of time. Reduced warehousing costs, higher customization, and feasibility make 3D printing the best fit for the defense industry.

7. *Weight Reduction:* Weight is one of the most important factors to consider when it comes to aircraft design. There are countless ways in which the military is currently working to make lighter weight and safer machinery, equipment and vehicles for use in combat as well as daily operations. It is not unusual for a single soldier to carry between 90 to 140 lbs worth of gear, including weapons, ammo, water, MREs, batteries, and personal protective equipment. Increased weight in transport vehicles, planes, and ships can decrease fuel efficiency and reduce maneuverability and speed. As a result, the military has a lot of interest in developing ways to lower weight without sacrificing performance. The US Army is investigating lightweight metals such as titanium, titanium alloys, and hybrid ceramic tile composites for their next-generation combat vehicles.

8. *Part Consolidation:* One of the key benefits of 3D printing is part consolidation: the ability to integrate multiple parts into a single component. Reducing the number of parts needed can significantly simplify the assembly and maintenance process by reducing the amount of time needed for assembly.

9. *Democratization:* The 3D printing technology has brought democratization. It highlights that manufacturing homemade, illegal weapons has become more accessible. People for a very long period of time have not been able to manufacture craft weapons. The emergence of 3D printing technology changes everything. People could not manufacture weapons before, and now they can do it due to the 3D printing.

5.6 CHALLENGES

Although military prototypes can be 3D printed quickly and cheaply, wider implementation of additive manufacturing for the production of end parts still faces a number of challenges. Challenges such as material limitations, standardization, and certification processes persist. As with any emerging technology, there are potential challenges related to security risks and vulnerabilities. Other challenges of 3D printing in defense industry include the following [8]:

1. *Quality assurance:* One of the main concerns for the defense industry is quality assurance, as all parts must adhere to stringent performance requirements. The military must be assured of the repeatability and accuracy of the 3D printing production process.

2. *No Standards*: Lack of standards also can be a problem. Speed of use and versatility sum up what makes additive manufacturing different. Currently there are no fully defined industry standards for 3D printed parts in the defense industry. It is crucial to establish a comprehensive set of standards to govern 3D printing processes and qualify printed parts. Materials must also be certified by a defined standard, increasing the developmental work required by DoD to leverage a

still immature field to this task. Safety standards must be published and implemented. There are currently unclear regulations that prevent 3D-printed weapons from reaching ordinary citizens.

3. *Skills Gap:* Although 3D printing has been used within the defense industry for a number of years, there remains an AM skills gap. Further training in the particularities of AM production, designing for AM, maintenance and supply chain management will be necessary to push forward the use wider use of additive manufacturing within the military. The US has been able to cultivate a strong academic foundation and attract global talent in AM, with the US commanding the largest share of the global industrial 3D printer market at 33%. While slightly trailing the US in academic impact, China has made significant strides in the commercial sector.

4. *Digital Security*: If the military is to use additive manufacturing for on-demand, localized production, it will be vital to ensure the security of the digital CAD files. This will require additional digital security measures to ensure files cannot be accessed externally, and digital supply chains remain secure. Ensuring the integrity of the 3D printing process and preventing the production of counterfeit components are crucial considerations in safeguarding national security interests.

5. *Security Threats:* 3D printing could enable new criminal and security threats, such as the ability to print weapons. 3D printing could allow weapons to enter circulation without registration, and printed plastic or ceramic guns could render common security measures ineffective. Western armed forces are concerned that adversaries could use 3D printing to copy or develop advanced weapon systems.

6. *Weapon Limitations:* 3D-printed guns made of plastic can be damaged by heat exposure and they have a limited capacity before needing to be cooled off. They can also be vulnerable to jamming and misalignment.

7. *Vulnerability to Cyber Attacks:* The digital transfer of files during 3D printing can expose sensitive information about the design or location of facilities. This can leave the system vulnerable to cyber attacks that could steal data or cause malfunction.

These challenges stand in the way of unleashing the full potential of 3D printing for military use.

5.7 CONCLUSION

It has been predicted that the defense industry will expand it's arms towards 3D printing in the next 10 years. 3D printing has the potential to fundamentally change how parts are manufactured and delivered. This may have enormous implications for the future of the military supply chain. In an increasingly complex military landscape, 3D printing has the potential to meet key defense needs. The use of 3D printing in the military enables them to be more self-sufficient. With 3D printers, militaries can print replacement parts and components for their equipment, rather than having to rely on outside suppliers.

3D printing technology is revolutionizing the military industry by providing a way to quickly and cheaply produce spare parts, prototypes, and even entire weapons systems. In the future, 3D printing technology will become even more sophisticated and widespread. It will have a transformative impact on the way militaries operate [11]. The future of 3D printing for global militaries certainly looks bright. More information about 3D printing technology in the defense industry can be found in the books [15-17].

REFERENCES

[1] C. Collins, "Additive manufacturing," October 2019, https://www.defensemedianetwork.com/stories/additive-manufacturing-department-of-defense-3d-printing-military-logistics/

[2] M. N. O. Sadiku, U. C. Chukwu, and J. O. Sadiku, "3D printing in defense," *Innovative Mult-disciplinary Journal of Applied Technology*, vol. 2, no. 5, 2024, pp. 226-237.

[3] F. R. Ishengoma and T. A. B. Mtaho, "3D printing: Developing countries perspectives computer engineering and applications," *International Journal of Computer Applications*, vol. 104, no. 11, October 2014, pp. 30-34.

[4] M. N. O. Sadiku, S. M. Musa, and O. S. Musa, "3D printing in the chemical industry," *Invention Journal of Research Technology in Engineering and Management,* vol. 2, no. 2, February 2018, pp. 24-26.

[5] D. Pitukcharoen, "3D printing booklet for beginners," https://www.metmuseum.org/-/media/files/blogs/digital-media/3dprintingbookletfor beginners.pdf

[6] C. Connolly, "The transformative power of 3D printing: From imagination to reality," May 2024, https://medium.com/@ciaranpconnolly/the-transformative-power-of-3d-printing-from-imagination-to-reality-a31d5e48fc3b#:~:text=3D%20printing%20has%20the%20power,up %20a%20world%20of%20 possibilities.

[7] "How is 3D printing transforming the defence industry?" June 2018, https://amfg.ai/2018/06/19/how-3d-printing-is-transforming-the-defence-industry/

[8] "Defence innovations with 3D printing military applications," November 2019, https://www.zeal3dprinting.com.au/defence-innovations-with-3d-printing-military-applications/

[9] T.M. Issa, A.A. English, "3D-printed weapons: Interpol and defense experts warn of 'serious' evolving threat," https://english.alarabiya.net/features/2022/11/01/3D-printed-weapons-Interpol-and-defense-experts-warn-of-serious-evolving-threat-

[10] C. Lee, "Army Gung-Ho on 3D printing spare parts," December 2020, https://www.nationaldefensemagazine.org/articles/2020/12/11/army-gung-ho-on-3d-printing-spare-parts

[11] C. Harris, "Best 3D printers for military applications," April 2023,
https://all3dp.com/1/3d-printing-military-applications/
https://all3dp.com/1/3d-printing-military-applications/#google vignette

[12] "Industrial applications of 3D printing: The ultimate guide,"
https://amfg.ai/industrial-applications-of-3d-printing-the-ultimate-guide/

[13] M. Roaten, "Military looks for novel ways to employ 3D printing," June 2021,
https://www.nationaldefensemagazine.org/articles/2021/6/11/military-looks-for-novel-ways-to-employ-3d-printing

[14] "The military turns to 3D printing," June 2023,
https://shop3duniverse.com/blogs/digital-fabrication-in-the-news/the-military-turns-to-3d-printing#:~:text=3D%20printing%20technology%20is%20being,turnaround%20times%20for%20design%20changes.

[15] A. B. Badiru, V. V. Valencia, and D. Liu (eds.) *Additive Manufacturing Handbook: Product Development For The Defense Industry*. Boca Raton, FL: CRC Press, 205.

[16] United States Government Accountability Office, *Defense Additive Manufacturing Dod Needs to Systematically Track Department-Wide 3D Printing Efforts*. CreateSpace Independent Publishing Platform, 205.

[17] G. Blokdyk, *3D Printing in Aerospace and Defense Standard Requirements*. Emereo Pty Limited, 2018.

CHAPTER 6

INTERNET OF THINGS IN THE MILITARY

"The Internet of things is creating a new world where everything is connected and can be controlled remotely."
– Tim Cook

6.1 INTRODUCTION

The US economy is made of complex and interconnected networks which connect numerous nodes, including corporations, financial institutions, academic institution, factories, and urban centers. These networks have features that make their behavior unstable and unpredictable. Terrorists can exploit these features. Modern information and communications technology, particularly the Internet, has become central to current modes of activism and politics around the world.

Every major corporation now uses web-based infrastructure, potentially vulnerable to criminals or spies. Our societies are wide-open targets for terrorists. There is increased vulnerability of the West's economic and technological systems. Homeland defense provides border protection and protection from threats against the populace. Such protection causes fewer civilians and combatants to die. Modern military operations are conducted in a complex, multidimensional, highly dynamic and disruptive, and unpredicted environment. Military commanders operate under strong time pressures and high operational tempos. Increase in anti-military activities lead to increased need of computer applications in military environments. Modern military and defence systems include drones, spacecrafts, missiles, military vehicles, ships, marine systems, satellites. and rockets. The Internet of things (IoT) can help the military solve some of its problems. By integrating systems of connected sensors, actuators, and other IoT devices into existing military infrastructures, the military can become more efficient and effective.

New military technologies, of which nanotechnology leads the way, can even change the entire nature of the world system and alter the battlefield, international relations, and society [1]. The Internet of things (IoT) has been gradually bringing a sea of technological changes in our daily lives. It is now regarded as the future of smart technologies. The IoT has been developed widely across the world with an emphasis on civil applications. Today, IoT is changing combat operations. The government of each nation find ways to utilize the IoT for military purposes. Modern military environments require effective and reliable technologies such as IoT. The IoT has strong military applications, connecting ships, planes, tanks, drones, soldiers, and operating bases in a cohesive network. The battlefield of the future will be densely populated by a variety of "things," including sensors, drones, weapons, vehicles, robots, and human-wearable devices. Sensors can scan irises, fingerprints, and other biometric data to identify individuals who might pose a danger. Many nations are investigating the potential military benefits of Internet of things technologies [2].

This chapter introduces how the defense industry can leverage the opportunities created by IoT. It begins with providing an overview on IoT. It presents military IoT. It covers some military applications of IoT devices. It highlights the benefits and challenges of military IoT. The last section concludes with comments.

6.2 OVERVIEW ON IOT

The Internet began with some military computers in the Pentagon called Arpanet in 1969. It expanded throughout the 1980s as a set of four parallel military networks, each at a different security level. The core technology which gives the Internet its particular characteristics is called Transmission Control Protocol/ Internet Protocol (TCP/IP), which is essentially a set of rules for communication [3].

Internet of things (IoT) is a worldwide network that connects devices to the Internet and to each other using wireless technology. IoT is expanding rapidly and it has been estimated that 50 billion devices will be connected to the Internet by 2020. These include smart phones, tablets, desktop computers, autonomous vehicles, refrigerators, toasters, thermostats, cameras, alarm systems, home appliances, insulin pumps, industrial machines, intelligent wheelchairs, wireless sensors, mobile robots, etc.

There are four main technologies that enable IoT [4]:

(1) Radio-frequency identification (RFID) and near-field communication.
(2) Optical tags and quick response codes: This is used for low cost tagging.
(3) Bluetooth low energy (BLE).
(4) Wireless sensor network: They are usually connected as wireless sensor networks to monitor physical properties in specific environments.

Other related technologies are cloud computing, machine learning, and big data. Figure 6.1 illustrates some applications of the Internet of things [5].

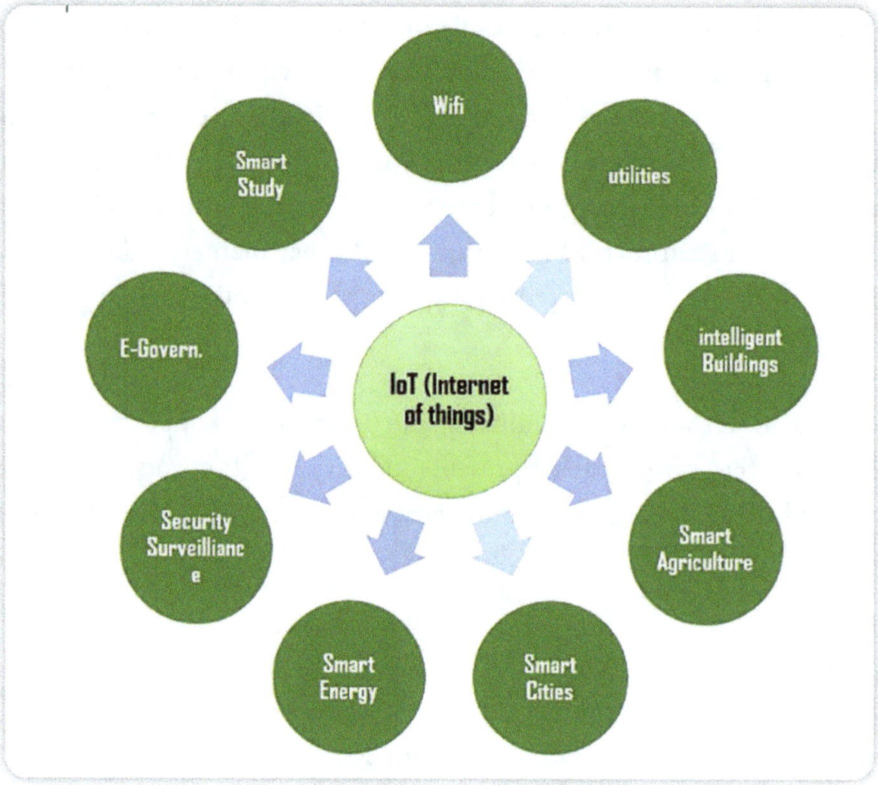

Figure 6.1 Some applications of Internet of things [5].

IoT technology enables people and objects to interact with each other. It is employed in many areas such as smart transportation, smart cities, smart energy, emergency services, healthcare, data security, industrial control, logistics, retails, structural health, traffic congestion, manufacturing, and waste management. The Internet of things is extensively developed world-wide with a focus on civilian applications such as electric power distribution, intelligent transportation, healthcare, industrial control, precision agriculture, environmental monitoring, etc. Military logistic is perhaps the most important application domain for IoT [6].

IoT supports many input-output devices such as camera, microphone, keyboard, speaker, displays, microcontrollers, and transceivers. It is the most promising trend in the healthcare industry. This rapidly proliferating collection of Internet-connected devices, including wearables, implants, skin sensors, smart scales, smart bandages, and home monitoring tools has the potential to connect patients and their providers in a unique way.

Today, smartphone acts as the main driver of IoT. The smartphone is provided with healthcare applications.

6.3 MILITARY INTERNET OF THINGS

The military has been an early adopter of IoT and has been investing heavily in IoT. The Pentagon sees the Internet as the "fifth domain" of warfare, alongside land, air, sea, and space [7]. The IoT has diverse military applications, connecting ships, planes, tanks, drones, and soldiers. One way the Department of

Defense (DoD) aims to speed up and automate decision-making is through a massive military Internet of Things (IoT) and artificial intelligence (AI). As shown in Figure 6.2, a massive military IoT promises a host of battlefield benefits in such areas as unmanned surveillance and targeting, situational awareness, soldier health monitoring, and other critical applications [8]. In these applications, the military is supposed to protect sensitive or important information involving names, addresses, social security numbers, medical records, financial records, etc. The measures we take to protect our information assets can generally be described in terms of the classic CIA triad of confidentiality, integrity and availability.

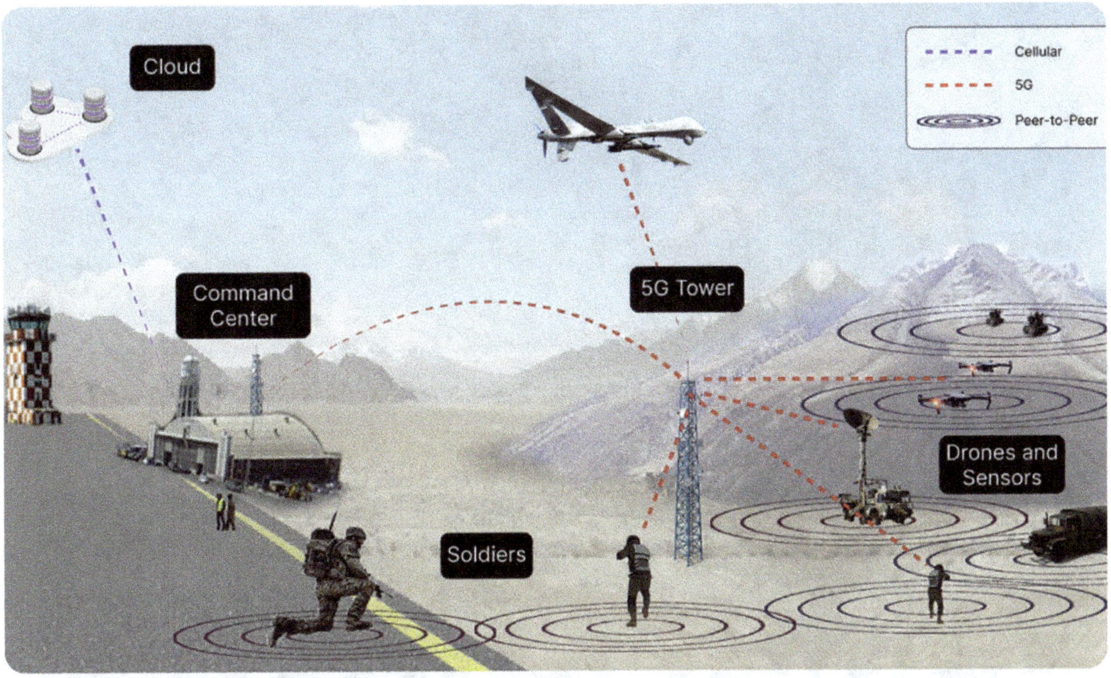

Figure 6.2 A massive military IoT promises a host of battlefield benefits [8].

The military Internet of things (MIoT) (also called Internet of Military Things (IoMT) or Internet of Battlefield Things (IoBT)) is an information system which can capture the physical attributes, state information of military people, equipment, and materials by several information sensing means. IoMT is a class of heterogeneous, complex network of various sensors, actuators, and devices found on the ground, sea and in space which provide information technology for combat operations, reconnaissance, and even warfare capabilities. Connected "things" include soldiers, equipment, vehicles, drones, robots, human wearable devices, biometric devices, weaponry, armor, a host of smart technologies, and all others assets Army needs to win a battle . Some of these devices are shown in Figure 6.3 [9]. It involves the full realization of pervasive sensing, pervasive computing, and pervasive communication providing a strategic advantage. To the extent that commanders can improve their awareness of the condition of their things, they will have a better understanding of how to optimize the force for a given mission. Figure 6.4 shows a typical military IoT [10].

Figure 6.3 Some of the connected devices in MIoT [9].

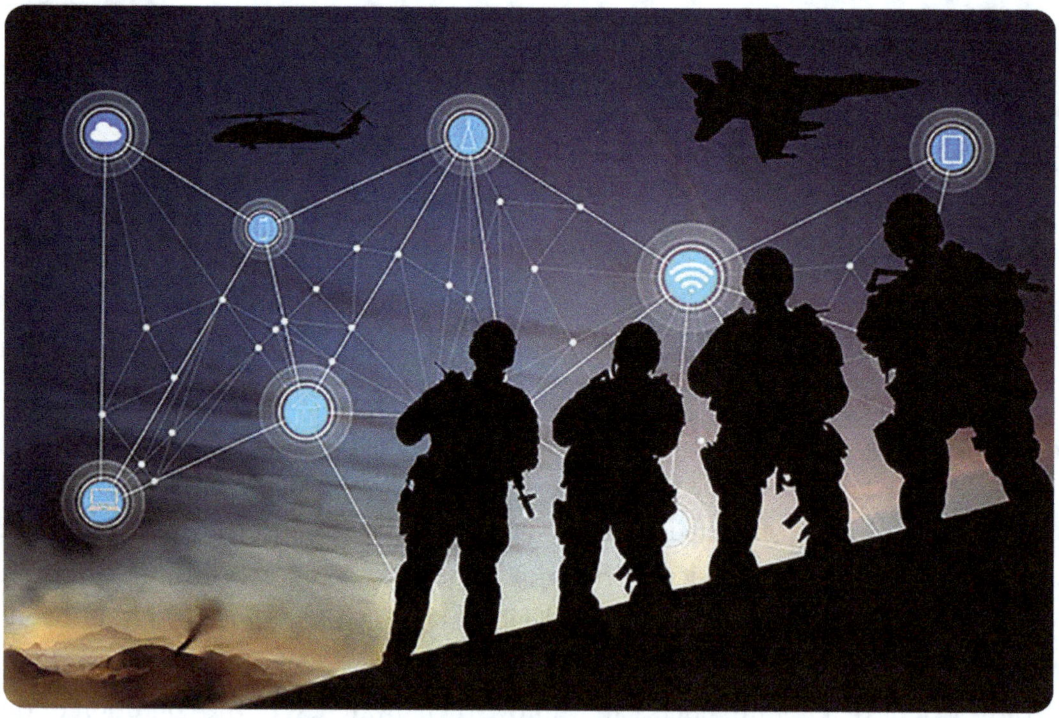

Figure 6.4 A typical military IoT [10].

Various sensing and computing devices are worn by soldiers and embedded in their combat suits, helmets, and weapons systems. To some extent, modern military equipment could be regarded as sensors or actuators and integrated into the rest of the military information infrastructure. There are different kinds of sensors used in military operations. Sensors are used in flight controls, environmental monitoring, weaponry controls, indicators, communications, security systems, explosives detection systems, chemical warfare, crime detection systems, intrusion detection systems, battlefield surveillance systems, manned and unmanned aerial platforms, camera, etc. These sensors have many features including user-friendly, easily accessible, flexible, self-testing, self-diagnosis, self-compensation, etc. [11].

6.4 MILITARY APPLICATIONS

Military applications of IoT devices are on the rise. The adoption of the IoT to military applications has a substantial impact on soldiers on the battlefield. IoT systems and equipment are being used more and more by military organizations to improve operational efficiency and security. IoT is a technology that can change any system when applied in a useful manner. IoT technology can be used to meet military needs such as base operations, situational awareness, medical care, energy management, and surveillance, A typical IoT application will be allied to a huge number of devices. A typical RFID based target sensing application is shown in Figure 6.5 [6]. Some applications of IoT in the defence and military fields include the following [12-14]:

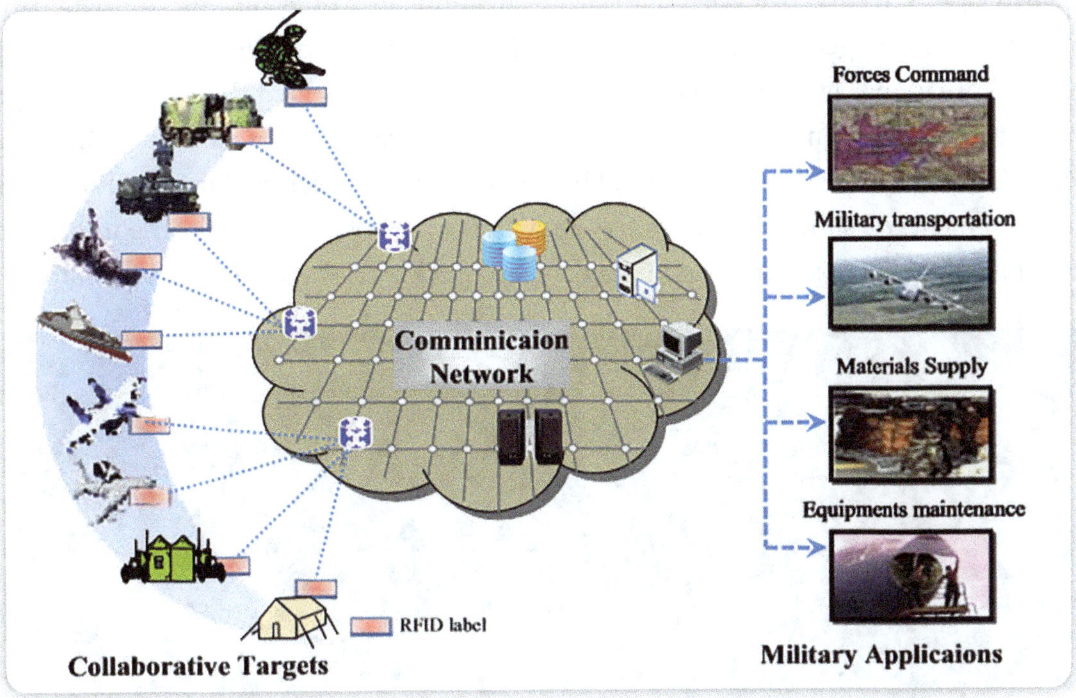

Figure 6.5 A typical RFID based target sensing application [6].

- *Autonomous Weapon Systems:* The military uses a wide range of sensors and unmanned vehicles for gathering intelligence. With the aid of robots, autonomous and smart detection of harmful chemicals and biological weapons is possible. There is also control of land vehicles and aircrafts, without neither the presence, nor the coordination of humans. Autonomous surveillance and weapons systems can bring precision to the battle via AI and technologies such as facial recognition that can target enemy combatants more accurately than humans. The distinctive feature these systems present is their ability to select and eliminate their targets without human intervention.

- *Smart Fort*: When city is struck by disaster, some armed force is often deployed to the city to provide relief. The sensors, actuators, effectors, and smart devices. deployed by the city administration provide valuable data for disaster recovery operation. Each soldier is equipped with a sensor system for health and equipment monitoring. While soldiers are usually protected by effective high technology body armor, additional safety is provided by the weapon system's connectivity to the network. The development of smart cities in the civilian sector could translate to Smart Forts for the Army. The Army has already begun to implement some elements of Smart City technology in water and energy management, but much more could be done to exploit the applications in security, traffic management, health care, and personnel readiness.

- *Smart Bases:* Incorporating IoT devices and sensors into military bases can have several positive effects. Smart management of resources – electricity and water for example – will increase the capacity and output of military bases while ensuring that the wellbeing of all individuals inside the base is protected.

- *Defence:* Today, military and homeland security equip its workforce with the basic functions provided by IoT technologies (e.g. RFID). IoT technologies offer promise for data gathering, as well as information generation and dissemination. IoT functionalities are useful for establishing advanced situational awareness in the area of operations. Augmented Reality (AR) engage millions around the world to find things in their own physical environment using a smartphone camera and an AR overlay. For the military, vital information can be projected using AR headsets already in development. As shown in Figure 6.6, imagine being able to see real-time information about the terrain, weather conditions and being able to quickly distinguish friend from foe [15].

Figure 6.6 A soldier using AR headsets [15].

- *Logistic Tracking:* Logistics is an area where multiple low-level sensors are already being used in defense. The military has already deployed some IoT technologies in non-combat scenarios. IoT technology can be used to track supplies and equipment from their source to where they are needed on the battlefield. For example, the deployment of RFID tags, sensors, and standardized barcodes allow for tracking individual supplies.

- *Data Warfare:* It is difficult to overemphasize the importance of information to military commanders everywhere. For any battle, information is the key to victory. This data or information will provide actionable intelligence, as shown in Figure 6.7 [16]. Since information is critical to all activities, the military is hungry for technology or tools that improve processing of information. The Internet of things (IoT) is one such technology. By collecting data from different military platforms – including aircraft, weapon systems, ground vehicles, and troops themselves – the military can increase the effectiveness of their intelligence and surveillance systems. This information will allow the military to identify key threats faster.

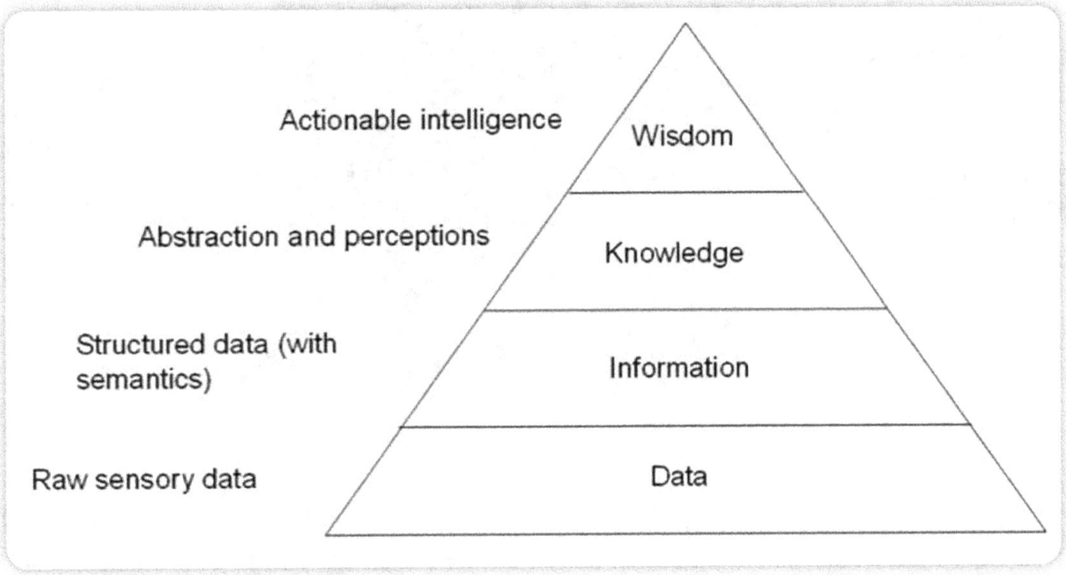

Figure 6.7 How information and data produce actionable intelligence [16].

- *Wearables:* From autonomous combat systems to wearables, the Internet of Military things is augmenting performance in the sector. The military wearables market is expanding rapidly, driven by continuous innovation in small, lightweight products that can be used for navigation, communication, and computing systems, vision and surveillance, and biometric monitoring. These products have typically been worn as headgear, eyewear, wristwear, or ear-worn devices. Often called the Internet of battlefield things, a network of intelligence-gathering and biometric body sensors embedded in soldiers' combat uniforms, helmets, weapons systems, and transports can convey valuable battlefield information together with soldier location. Typical military wearables are shown in Figure 6.8 [17]. One of the challenges of future soldier implementation is related to the dimensions of the wearable tech itself.

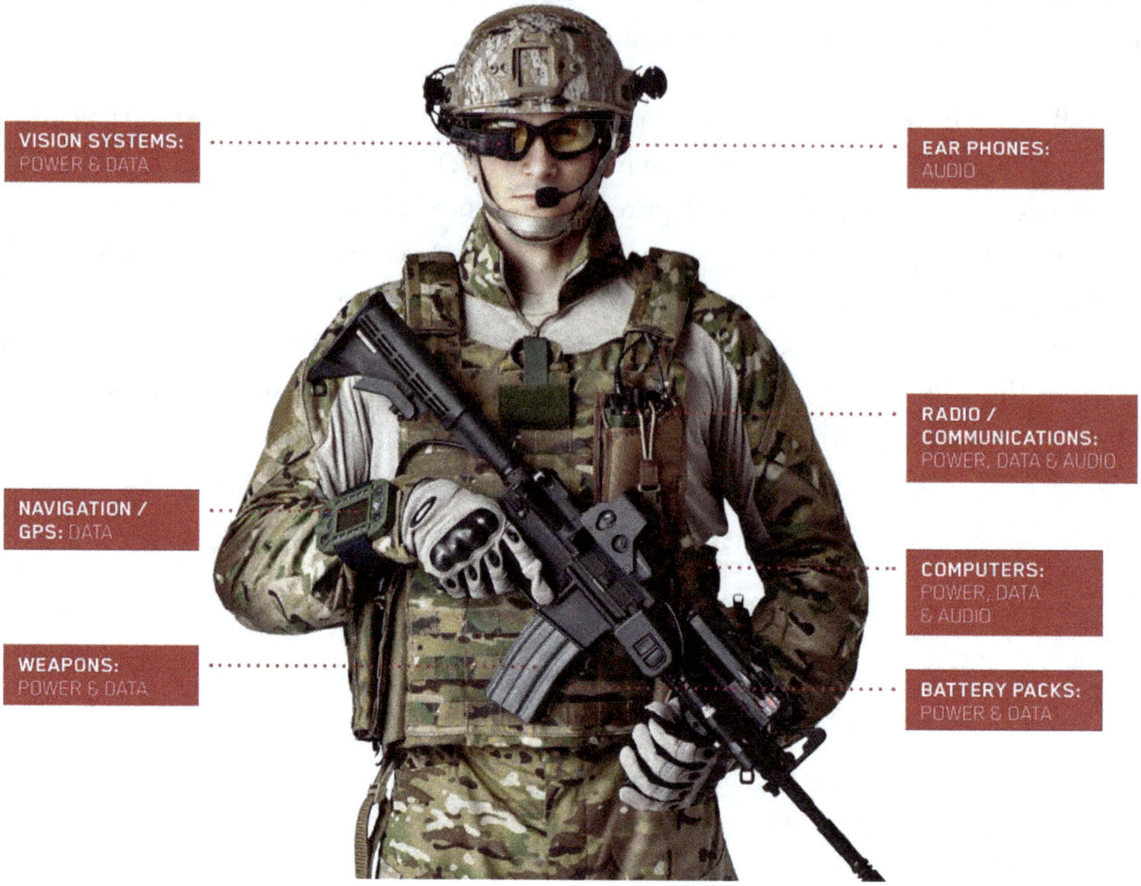

Figure 6.8 Typical military wearables [17].

- *Monitoring Soldier Health:* It is quite difficult to monitor the health of a soldier on the field. A large range of sensors can be attached to the soldiers' jacket, which can track, sense, and send alerts about its changing medical conditions to the Command Center. Through the use of sensors in military uniforms, a soldier's heart rate, body temperature, and thermal distribution can be monitored. The collected data about a soldier's physical and mental health can then be shared with doctors in real-time, allowing for the arrangement of any necessary aid in advance on a per-soldier basis.

- *Drones:* Defense forces worldwide employ drones for surveillance, combat missions, and target decoys. IoT enables armed forces to survey the battlefield with unmanned aerial drones that are equipped with cameras and sensors. These drones can transmit real-time data to the command center, capture live images, and track the terrain and locations of the enemies. A swarm of military drones is shown in Figure 6.9 [18]. Swarming is a method of operations where multiple autonomous systems act as a cohesive unit by actively coordinating their actions. An object recognition used by a drone recognizes the vehicles, persons, and monitors their movements from an altitude. These data are collected and analyzed in real time, which gives the army staff the edge before they go to the attack. The US military also uses drones for the independent patrol near borders to alert military staff in the event of an intruder.

Figure 6.9 A swarm of military drones[18].

- *Data Warfare:* By collecting data from a wide range of military platforms – including aircraft, weapon systems, ground vehicles, and troops themselves – the military can increase the effectiveness of their intelligence, surveillance, and reconnaissance systems. This wealth of information will allow the armed forces to identify key threats faster and with more accuracy.

6.5 BENEFITS

The usage of IoT in the military and defense has become essential due to the rise in anti-military actions. IoT is used in military environment such as smart bases that incorporate commercial IoT technologies in facilities. The IoT provides a number of military benefits, including lower costs for technology, greater awareness of the combat situation, and effective decision making. IoT has the potential to change how soldiers in the field shoot, move, and communicate. Many nations in military and defense applications are seeking to use the IoT as a means of addressing various issues in war and fighting. Military research on IoT is essentially about daily work in national defense such as military logistic, whose application mode is similar to it in civil domain. When cloud connections are feasible, IoT can take advantage of the cloud's massive scalability and processing power. Other benefits of IoT in the military include the following [9,19]:

- *Efficiency:* IoT is largely about making people live and work smarter. IoT devices will aid in combat by increasing the efficiency of military operations and surveillance. IoT devices can be used to aid the military in all aspects of their operations. The military may become more efficient and effective by integrating sensor systems, actuators, and control systems with current military infrastructures.

- *Communication:* Lack of communications in military operations may result in significantly high casualties. The ability for leaders to communicate with soldiers – even at extreme distances – has of greatly improved in the past years due to military IoT. IoT can be used to improve logistics and communications. Commanders can make informed decisions with real-time analysis of the data. To collect data, they can use sensors, cameras, unmanned vehicles and soldiers, as well as sensors and cameras on ground. Data fusion and operations communications can give a complete picture of the battlefield.

- *Automation:* Saving the lives of soldiers is a major concern. This problem can be eliminated by the means of automation. Detection of enemy and their elimination can be automated with the help of wireless sensor networks.

- *Battlefield Surveillance:* Although human beings continue to be the principal battlefield agents and drivers of success in autonomous weapons system, autonomous surveillance and weapons systems such as military drones, smart missiles, and unmanned ground vehicles can conduct advanced battlefield surveillance, enhance battle intelligence, and even engage targets to preserve soldiers' lives.

- *Soldier's Health:* Knowing a soldier's health condition is another use of IoT in defense and the military. Certain military applications can improve combat operations by monitoring soldiers' health with sensors. This includes monitoring heart rate, body temperature, and thermal distribution.

- *Water Management:* IoT devices monitor many systems including water, power and energy consumption. IoT can be used to help military water management.

- *Competitive Advantage*: The ability of the Army to understand, predict, adapt, and exploit the vast array of connected devices that will be present of the future battlefield is critical to maintaining and increasing its competitive advantage.

6.6 CHALLENGES

There are challenges with IoT military adoption. IoT brings with it security challenges that present both opportunities and obstacles. The digital exchange of information across national borders may be a major challenge. Civilian mobile telecommunication such as 4G LTE will need to be modified for advanced, military communications architectures. For the military, there is no one-size-fits-all solution to the IoT. One must keep in mind that incorporating IoT technology can be something of a cautionary tale. Although MIoT projects abound, some of them are uncontroversial and have little to do with warfare.

The IoT technologies will give our enemies ever increasing capabilities that must be countered. Tradeoffs in successful IoMT implementation will exist between interoperability, seamless information sharing, decision-making, and opening up the cybersecurity threat landscape. Other challenges of military IoT include the following [20]:

- *People:* What is likely to be the weakest link in our defense: people. Deciding on the division between human and autonomous decision-making will be one of the big moral and technical challenges linked to the success of autonomous systems.

- *Commanders:* As the technological capabilities of allies and adversaries advance, more pressure is put on military commanders to anticipate, assess, and take action in increasingly pressurized environments. Military commanders work at a solid time and high rhythms of operation. The time span of officers is constantly shorter for acquiring a precise assessment, surveying, and deciding possible gaming strategies.

- *Ethical Use of Technology:* Military technologies have always been fraught with ethical concerns. For example, while autonomous weapon systems have been gaining much consideration among Armed Forces of a number of nations, they are also becoming an immediate cause of humanitarian, ethical, and legal concern. The ethical use of technology in warfare needs must be consistent with doctrines governing the ethics of war.

- *Diversity:* The sheer number of devices (sensors, drones, and computing devices) could lead to information overload, more cyber vulnerabilities, and network management issues. Diversity of data poses unique challenges to cloud implementation in the military.

- *Vulnerability:* The immense promise of the military IoT comes with immense risks. While there have always been risks to DoD sensors and controls, their proprietary nature and isolation limited the possibility of attack. Now, with such capabilities being given Internet access, DoD is entering a quickly deepening pool of vulnerability. At risk are all the things that embrace the Internet of Things (IoT): DoD facilities, equipment, employees, and their possessions—any of which could be used to cause harm. We could soon be in a position where a determined adversary could shut down our power and water, turn off our security systems, disrupt our ability to provide medical care, listen to our conversations, and monitor our movements.

6.7 CONCLUSION

Military IoT includes everything from battlefield sensors and weapons systems to tracking devices, communications equipment, wearables, drones, ships, planes, tanks, and even body sensors. Military IoT adoption is still in its infancy. Nothing differentiates MIoT from IoT more than the military. Although military application of IoT is increasing in military, it faces the same challenges as the commercial IoT, such as standardization, scalability, interoperability, and security [13]. To become a reality, MIoT will have to overcome these major challenges. The future of military combat is going high-tech as scientists create an Internet of things for combat gear embedded with biometric wearables and other connected devices to help soldiers identify the enemy and perform better in battle.

Many countries have actively invested in IoMT. The US military has developed an integrated warfighting network that converges and combines all the data from IoMT sensors, radars, and satellites. Like the US, China has also developed a strategic outline for integrated warfare. The Indian Army is using IoMT for communication purposes. The IoT ecosystem in Pakistan is nascent as the country lacks the basic infrastructure to produce IoT devices on a large scale [9]. More information on MIoT can be found in the books in [1,21-24].

REFERENCES

[1] F. Allhoff, P. Lin, and D. Moore, "Military," in *What Is Nanotechnology and Why Does It Matter? From Science to Ethics.* Wiley-Blackwell, chapter 9, 206.

[2] M. N. O. Sadiku, E. Awada, and S. M. Musa, "Military Internet of things," submitted to a journal.

[3] M. Townes, "The spread of TCP/IP: How the Internet became the Internet," *Millennium: Journal of International Studies*, vol. 41, no. 1, 2012, pp. 43 –64.

[4] M. N. O. Sadiku, and S. M. Musa and S. R. Nelatury, "Internet of things: An Introduction," *International Journal of Engineering Research and Advanced Technology*, vol. 2, no.3, March 2016, pp. 39-43.

[5] "Internet of things (IoT),"
https://www.geeksforgeeks.org/internet-things-iot-2/

[6] L. Yushi, J. Fei, and Y. Hui, "Study on application modes of military Internet of things (MIOT)," *IEEE International Conference on Computer Science and Automation Engineering,* May 2012, pp. 630-634.

[7] W. Fripp, "@War: The rise of the military-internet complex," Intelligence and National Security, vol. 31, no. 5, 2016, pp. 783-785.

[8] A. Fish, "IoT, AI, and the future battlefield," September 2022,
https://militaryembedded.com/ai/deep-learning/iot-ai-and-the-future-battlefield

[9] M. Shafeeq, "Internet of military things (IoMT) and the future of warfare," December 2022,
http://geostrategicmedia.com/2022/12/04/internet-of-military-things-iomt-and-the-future-of-warfare/#:~:text=The%20Internet%20of%20Military%20Things%20%28IoMT%29%20is%20a,between%20unmanned%20vehicles%20and%20a%20central%20command%20station.

[10] "IoT for military: How the Internet of things can benefit the military"
https://www.iotworlds.com/iot-for-military-how-the-internet-of-things-can-benefit-the-military/

[11] "Modern sensors for defence and military applications," August 2018, https://electronicsforu.com/market-verticals/aerospace-defence/modern-sensors-defence-military-applications

[12] F. T. Johnsen, "Application of IoT in military operations in a smart city," *Proceedings of International Conference on Military Communications and Information Systems*, May 2018.

[13] P. Fraga-Lamas et al., "A review on internet of things for defense and public safety," *Sensors*, vol. 16, 2016.

[14] "The military benefits and risks of the Internet of things," April 2019, https://asb.army.mil/Portals/105/Reports/2010s/2016%20A%20The%20Military%20Benefits%20and%20Risks%20of%20the%20Internet%20of%20Things%20Report.pdf?ver=B8FJkGnH43LJVsa0X9py9A%3d%3d

[15] "Four ways IoT and space will revolutionize military ops," December 2022, https://www.lockheedmartin.com/en-us/news/features/2022/four-ways-iot-and-space-will-revolutionize-military-ops.html

[16] N. Suri and M. Tortonesi, "Military Internet of things (IoT), Autonomy, and things to come," https://static1.squarespace.com/static/53bad224e4b013a11d687e40/t/57e41eac8419c2f2791befb5/1474567857321/Panel+-+Military+IoT%2C+Autonomy%2C+and+Things+to+Come.pdf

[17] "Military wearables trends require connectors to scale down," August 2020, https://connectorsupplier.com/military-wearables-trends-require-connectors-to-scale-down/

[18] "Army advances learning capabilities of drone swarms," August 2020, https://www.army.mil/article/237978/army advances learning capabilities of drone swarms

[19] P. Suciu, "The Internet of military things could change the military landscape," December 2021, https://news.clearancejobs.com/2021/12/29/the-internet-of-military-things-could-change-the-military-landscape/

[20] "DoD policy recommendations for the Internet of things (IoT)," December 2016 https://dodcio.defense.gov/Portals/0/Documents/Announcement/DoD%20Policy%20Recommendations%20for%20Internet%20of%20Things%20-%20White%20Paper.pdf?ver=2017-01-26-152811-440

[21] D. E. Zheng and W. A. Carter, *Leveraging the Internet of Things for a More Efficient and Effective Military.* Center for Strategic & International Studies, September 2015.

[22] S. Harris, *@War: The Rise of the Military-Internet Complex.* New York: Dolan Book, 2014.

[23] J.M. Thomas, *The Military Internet of Things: Adapting Commercial Capabilities.* Air University, 2021.

[24] A. Swami et al. (eds.), *IoT for Defense and National Security.* Wiley, 2023.

CHAPTER 7

BIG DATA IN THE MILITARY

"Every company has big data in its future and
every company will eventually be in the data business."
– Thomas H. Davenport

7.1 INTRODUCTION

There is no doubt that information is the most precious commodity for any business. Data is information in raw format, while information represents data after processing and analysis. There is data everywhere. It has invaded all aspects of our life. Data comes from a variety of sources such as sensors, social media sites, smart phones, Internet, emails, ecommerce transactions, weather data, medical records, insurance records, RFID devices, video sharing, etc. This huge amount of data is collectively called big data. Big data refers to massive amount of data that are so large that traditional processing tools cannot cope. It is a high-volume, high-velocity, and high-variety information that requires special information processing tools. Because of these characteristics, big data requires new technologies and techniques to capture, store, and analyze. The cloud word for big data is shown in Figure 7.1 [1].

Figure 7.1 The cloud word for big data [1].

Our lives revolve around huge data sets. With the advent of various social media platforms and multinational companies, the generation of data has increased drastically. As its name implies, big data is a structured, semi-structured, and unstructured data, which is very big, fast, and comes in many forms. Big data may be regarded as a phenomenon since we can observe its effects like growing volume and variety. It has become the fuel that every industry needs today to flourish. Data is constantly generated in the modern era by everything. Vast amounts of data are generated by various sources like satellites, drones, social media, and sensors. Typical sources of big data are shown in Figure 7.2 [2].

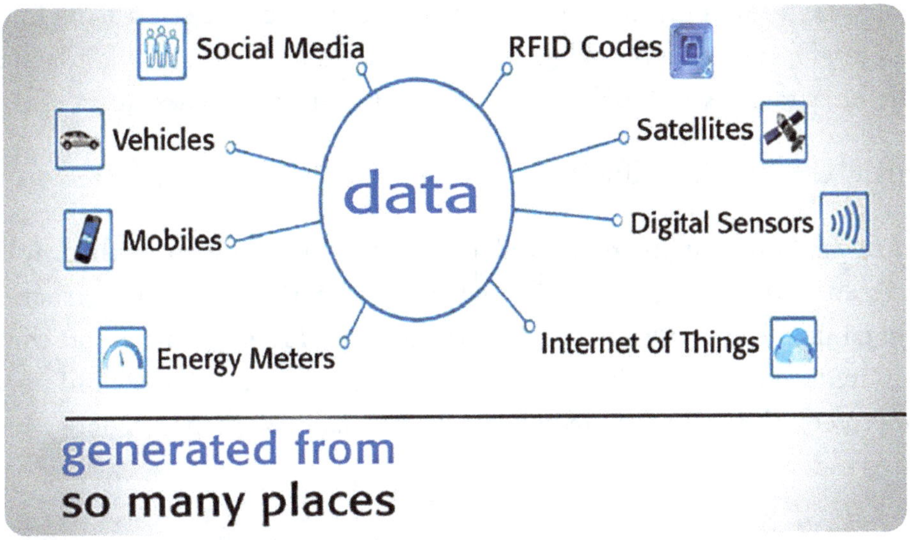

Figure 7.2 Typical sources of big data [2].

Large volumes of data, managed properly, are a boon for many industries, including the military. The US military is currently witnessing a significant shift in warfare, predominantly propelled by advancements in technology. At the heart of this paradigm shift lies the capacity to effectively gather, analyze, and rapidly and securely distribute essential information to military units. The current course of warfare has already started to value information superiority over mere firepower. The foundational principles governing battlefield dynamics, encompassing observation, engagement, mobility, communication, protection, and logistical support, are undergoing notable transformations. Although the prospect of big data in revolutionizing the battlefield is promising, its effective integration into defense intelligence analysis poses a series of challenges and opportunities [3]. Big data will play a key role in how the Army operates and wins its future combats.

Big data refers to massive volumes of structured and unstructured data that cannot be processed using traditional methods and is characterized by high volume, velocity, variety, veracity, and value. Data has become a crucial asset in various domains, including the military. The potential of big data to revolutionize battlefield dynamics has attracted considerable attention within military and intelligence circles. Big data has become a vital weapon system in modern warfare. It will determine the results of current and future conflicts [4].

In this chapter, we will delve into the role of big data in military operations and explore the realm of possibilities in military data analytics. The chapter begins with explaining what big data is all about. It covers the characteristics of big data. It discusses military data and provides some of its applications. It addresses military big data around the world. It highlights the benefits and challenges of military big data. It concludes with comments.

7.2 WHAT IS BIG DATA?

Big data applies to data sets of extreme size (e.g. exabytes, zettabytes) which are beyond the capability of the commonly used software tools. It involves situation where very large data sets are big in volume, velocity, veracity, and variability [5]. The data is too big, too fast, or does not fit the regular database architecture. It may require different strategies and tools for profiling, measurement, assessment, and processing.

Big Data is essentially classified into three types [6]:

- *Structured Data:* This is highly organized and is the easiest to work with. Any data that can be stored, accessed, and processed in the form of fixed format is known as a structured data. It may be stored in tabular format. Due to their nature, it is easy for programs to sort through and collect data. Structured data has quantitative data such as age, contact, address, billing, expenses, credit card numbers, etc. Data that is stored in a relational database management system is an example of structured data.

- *Unstructured Data:* This refers to unorganized data such as video files, log files, audio files, and image files. Any data with unknown form or the structure is classified as unstructured data. Almost everything generated by a computer is unstructured data. It takes a lot of time and effort required to make unstructured data readable. Examples of unstructured data include Metadata, Twitter tweets, and other social media posts.

- *Semi-structured Data:* This falls somewhere between structured data and unstructured data, i.e., both forms of data are present. Semi-structured data can be inherited such as location, time, email address, or device ID stamp.

The different types of big data are depicted in Figure 7.3 [7].

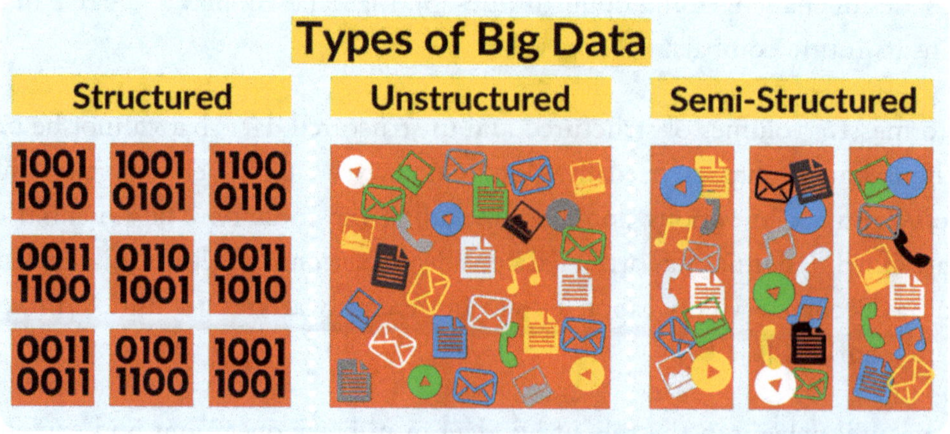

Figure 7.3 Types of big data [7].

One of the major strengths of big data is its flexibility and universal applicability to so many industries. Big data is used in several areas such as education, business, finance, government, healthcare, engineering, manufacturing, agriculture, social media, tourism, industry, entertainment, sports, construction, transportation, defense, etc. Increasingly, big data is regarded as the most strategic resource of the 21st century, similar in importance to gold and oil. It may also be regarded as the new form of currency [8].

The process of examining big data is often referred to big data analytics. It is an emerging field since massive computing capabilities have been made available by e-infrastructures. Analytics include statistical models and other methods that are aimed at creating empirical predictions. Data-driven organizations use analytics to guide decisions at all levels. Several techniques have been proposed for analyzing big data. These include the HACE theorem, cloud computing, Hadoop, and MapReduce [9].

7.3 CHARACTERISTICS OF BIG DATA

Big data is growing rapidly and expanding in all science and engineering, including physical, biological, and medical services. Different companies use different means to maintain their big data. As shown in Figure 7. 4 [10], big data is characterized by 42 Vs. The first five Vs are volume, velocity, variety, veracity, and value [2].

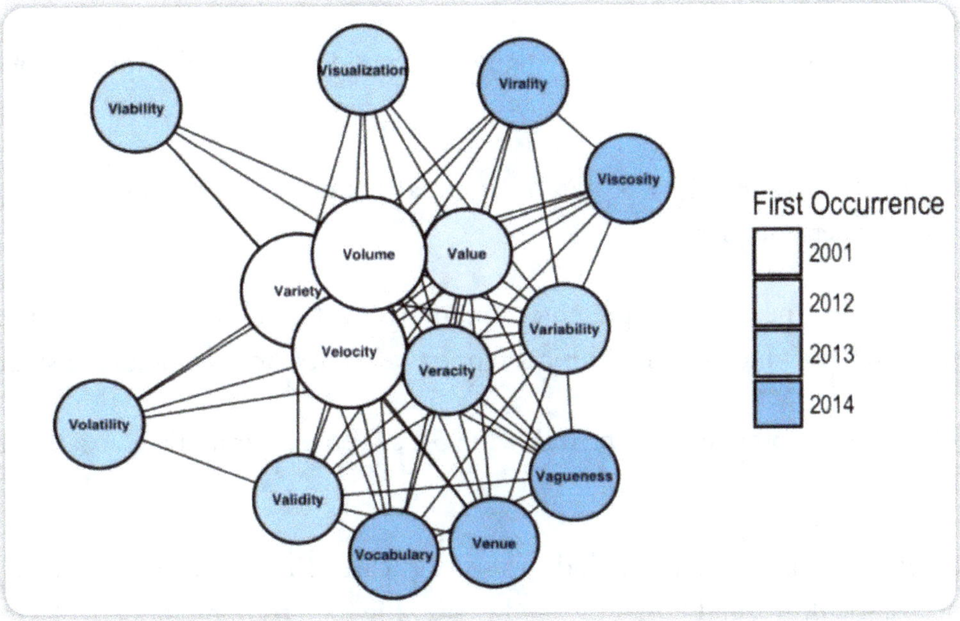

Figure 7.4 The 42 V's of big data [10].

- *Volume*: This refers to the size of the data being generated both inside and outside organizations and is increasing annually. Some regard big data as data over one petabyte in volume.

- *Velocity*: This depicts the unprecedented speed at which data are generated by Internet users, mobile users, social media, etc. Data are generated and processed in a fast way to extract useful, relevant information. Big data could be analyzed in real time, and it has movement and velocity.

- *Variety*: This refers to the data types since big data may originate from heterogeneous sources and is in different formats (e.g., videos, images, audio, text, logs). BD comprises of structured, semi-structured or unstructured data.

- *Veracity*: By this, we mean the truthfulness of data, i.e. weather the data comes from a reputable, trustworthy, authentic, and accountable source. It suggests the inconsistency in the quality of different sources of big data. The data may not be 100% correct.

- *Value*: This is the most important aspect of the big data. It is the desired outcome of big data processing. It refers to the process of discovering hidden values from large datasets. It denotes the value derived from the analysis of the existing data. If one cannot extract some business value from the data, there is no use managing and storing it.

On this basis, small data can be regarded as having low volume, low velocity, low variety, low veracity, and low value. Additional five Vs has been added [11]:

- *Validity:* This refers to the accuracy and correctness of data. It also indicates how up to date it is.

- *Viability:* This identifies the relevancy of data for each use case. Relevancy of data is required to maintain the desired and accurate outcome through analytical and predictive measures.

- *Volatility:* Since data are generated and change at a rapid rate, volatility determines how quickly data change.

- *Vulnerability:* The vulnerability of data is essential because privacy and security are of utmost importance for personal data.

- *Visualization:* Data needs to be presented unambiguously and attractively to the user. Proper visualization of large and complex clinical reports helps in finding valuable insights.

Instead of the 10V's above, some suggest the following 5V's: Venue, Variability, Vocabulary, Vagueness, and Validity) [12].

Industries that benefit from big data include the healthcare, financial, airline, travel, restaurants, automobile, sports, agriculture, and hospitality industries. Big data technologies are playing an essential role in farming: machines are equipped with sensors that measure data in their environment. Structured and unstructured data are generated in various types [13-15].

7.4 MILITARY BIG DATA

Big data represents a paradigm shift in data management, requiring innovative approaches to harness the potential value hidden within vast and complex data repositories. The advent of big data has revolutionized how armed forces conduct operations, enhancing their capabilities and effectiveness. By leveraging advanced analytics techniques, such as machine learning and artificial intelligence, militaries can extract actionable insights from vast datasets, enabling them to gain a competitive edge on the battlefield. Although the essence of war remains constant, its manifestations evolve over time. These evolving manifestations necessitate the adaptation of military methodologies. Figure 7.5 shows that using data is critical to the success on the battlefield [16].

Figure 7.5 Using data is critical to the success on the battlefield [16].

Defense and security have often been at the forefront of new technologies, but has lagged other industries with respect to data analytics. Militaries, in the current environment, use big data for surveillance, processing of intelligence, prioritizing of targets, engagement, post-strike damage assessment. Big data enables military planners to identify trends, patterns, and anomalies. It enhances situational awareness on the battlefield. In military war planning, big data analytics plays a vital role in converting raw data into actionable intelligence. Imagine soldiers equipped with real-time insights on enemy movements, logistics officers predicting supply chain bottlenecks before they occur. This is not science fiction; it is the power of data analytics in defense, as displayed in Figure 7.6 [17].

Figure 7.6 Soldiers equipped with real-time insights on enemy movements [17].

7.5 APPLICATIONS OF MILITARY BIG DATA

Big data offers diverse applications in military operations, including intelligence gathering, predictive maintenance, strategic decision-making, predictive analysis, situational awareness enhancement, threat assessment, decision-support mechanisms, surveillance, leadership, and optimization of logistical operations. Big data is being used in the military in the following specific ways [16,18]:

1. *Intelligence Gathering*: Intelligence is at the heart of all defense planning and implementation. One of the most critical aspects of military operations is intelligence gathering, which can now be augmented and improved using big data. Traditional intelligence gathering in the field includes teams splitting up, gathering information, returning to base, and writing reports, and then the different teams may or may not learn of what the other teams discovered. This is quickly becoming an outdated way to collect information. By analyzing massive datasets, big data can help identify patterns and trends in enemy behavior, troop movement, and communication networks. These patterns can provide valuable insights into enemy strategies, weaknesses, and potential threats. Big data analytics can also process and analyze satellite images and geographic data to provide an accurate and updated overview of the terrain, infrastructure, and resources in a conflict zone. This geospatial intelligence aids in planning and executing military operations with greater precision. In today's interconnected world, social media has become a goldmine of information. By using big data analytics, military forces can monitor and analyze the chatter on social media platforms to gather intelligence on public sentiment, enemy propaganda, and potential security threats. Figure 7.7 indicates that the future of defense is big data and military intelligence [18].

Figure 7.7 The future of defense is big data and military intelligence [18].

2. *Predictive Maintenance*: The maintenance and upkeep of military assets like vehicles, aircraft, and weaponry are essential for ensuring the operational readiness of armed forces. Big data analytics can help in predictive maintenance and resource optimization in many ways. First, by analyzing data from sensors and other monitoring devices, big data can identify early signs of wear and tear, potential system failures, and other maintenance issues in military equipment. This early warning system enables armed forces to address issues proactively, ensuring optimal performance and reducing the risk of unexpected breakdowns during operations. Second, big data analytics can help military forces optimize their resources by identifying inefficiencies and redundancies in their supply chains and logistics networks. By streamlining processes and allocating resources more effectively, armed forces can reduce costs, improve response times, and enhance their overall operational efficiency. Third, beyond equipment maintenance, big data can also be used to monitor the health and well-being of military personnel. By analyzing factors such as physical fitness, stress levels, and cognitive performance, big data can identify potential risks to soldiers' health and recommend preventative measures to maintain peak performance.

3. *Strategic Decision-Making:* The emergence of big data has prompted decision-makers and commanders to acknowledge its importance in shaping strategic insights and intelligence analysis. It marks the commencement of a new stage in data-centric decision-making. Big data has a transformative impact on strategic decision-making in military operations. By providing real-time insights and predictive analysis, it can help military leaders make informed decisions and improve the overall effectiveness of their strategies. Big data analytics can integrate data from multiple sources, including satellite imagery, surveillance feeds, and intelligence reports, to provide a comprehensive and real-time understanding of the operational environment. This enhanced situational awareness can help military commanders make better-informed decisions and adapt their strategies to changing circumstances. Big data can be used to create sophisticated war-gaming scenarios and simulations that allow military leaders to test and refine their strategies in a controlled environment. By analyzing the results of these simulations, military planners can identify potential weaknesses, anticipate enemy responses, and develop more effective strategies

for real-world operations. Big data can also be integrated into decision support systems that use artificial intelligence (AI) and machine learning algorithms to analyze and process information, providing military leaders with data-driven recommendations. These systems can help identify the most effective course of action, taking into account multiple factors such as mission objectives, available resources, and potential risks.

4. *Data-driven Leadership:* Leaders at all levels must understand data, data collection, data analysis, and what it means to the future of the Army. Giving leaders at all levels the understanding and tools necessary to navigate the data-centric environment allows for a more effective decision-making process. All leaders within the officer, warrant officer, and noncommissioned officer cohorts must develop valuable training for their formations if the Army is to be successful and take full advantage of the tools at its disposal. When leaders take the big data movement and apply artificial intelligence, they can discover trends, break through a typical bureaucratic structure, and make quick decisions with the right information. Predictive analytics plays a crucial role in cultivating data-driven leadership. Utilizing advanced analytical techniques to examine historical sustainment patterns with increased efficiency at the grassroots level offers significant advantages for future strategic planning efforts. This proactive approach assists leaders in better preparing their units to address a range of potential scenarios. Equipping leaders with suitable analytical instruments reveals opportunities to identify areas where resources are either underutilized or overstretched, such as in inventory stockpiles and maintenance facilities. Armed with this understanding, leaders can deploy resources more prudently, ensuring their distribution corresponds with operational demands. The impact of big data is particularly evident in addressing questions pertaining to who, what, where, and when, primarily with structured data. This underscores the crucial role of expert leadership and analysts in navigating intricate defense intelligence challenges. Figure 7.8 shows President Obama holding a meeting with military leadership [19].

Figure 7.8 President Obama holding a meeting with military leadership [19].

5. *Data Collection and Analysis:* This is integral to modern military operations. Data are emerging as a key component of military operations, both on and off the battlefield. The utilization of big data and advanced analytics provides militaries with unprecedented capabilities in war planning. By harnessing the power of data, military leaders can enhance situational awareness, predict enemy behavior, optimize logistics, and support decision-making processes. As the defense industry grapples with exponential increases in mission, data must be collected, managed, and analyzed. The promise of artificial intelligence (AI) to harness the power of big data and drive strategic decision-making represents an unprecedented shift in the industry. The collection and use of large data sets in weapons systems and for other purposes raise significant questions about data acquisition, retention, and privacy as well as bias. Data analytics can assist the Army in optimizing the allocation of limited resources. By giving leaders the proper tools, they can identify areas where resources are underutilized or overextended. Automated data collation and analytics would both save analyst effort and enable powerful new capabilities.

6. *Surveillance:* The fundamental role of big data in defense intelligence analysis revolves around the realm of military operations and surveillance. In the field of surveillance, it would be essential to know the areas to focus and the likelihood of attaining the best results with devices. Today, surveillance includes a plethora of devices. These are satellites, unmanned aerial vehicles, aerostats, Airborne Warning and Control System (AWACS), reconnaissance helicopters and aircraft, ground-based electronic devices, and human intelligence. Big data is used to develop sophisticated surveillance systems that enable real-time monitoring, threat detection, and early warning capabilities. For example, the United States ARGUS ground surveillance system mounted on a UAV collects more than 40 gigabytes of information per second. The data collected is massive. The information gathered through military operations can be put to varied use. Another prime example would be the US Air Force E-8 Joint STARS (Surveillance Target Attack Radar System) which conducts airborne ground surveillance; collects data on enemy positions, vehicles, and aircraft, collects imagery, and relays tactical pictures to ground and air theater commanders.

7.6 MILITARY BIG DATA AROUND THE WORLD

Military organizations around the world are often huge producers and consumers of big data. They stand to gain from the many benefits associated with data analytics. Worldwide, the amount of data gathered by the military grows, as does the desire and need by the military to extract and use this data to form actionable intelligence. Successful gathering, processing, and analyzing will effectively change warfare as it is understood today. We consider how the following nations are employing big data in their military operations.

- *United States:* The US is the top country for big data roles in the military industry. In the United States today, all aircraft already emit their locations through a system called automatic dependent surveillance–broadcast, and most ships do the same through the automatic identification system.

US warfighters operate in a more technologically augmented arena since September 11, 2001, where sensors, wearable computers, Internet of things (IoT)-enabled devices, and artificial intelligence (AI) systems all contribute to mission success. Today's warfighter often operates in remote, environmentally hostile, and actively contested regions. Their operations are increasingly dependent on analyzing data quickly to make critical decisions and respond to potential threats. Year after year, the US military faces increased operational commitments and budget constraints, thereby forcing it to do more with fewer resources. The National Security Agency (NSA) typically gathers SIGINT (Signals Intelligence) on terrorists, organizations, and persons with international or foreign associations using various methods. SIGINT is information about the actions, objectives, and capabilities of a foreign target acquired through the interception of signals and transmissions. A potential use of SIGINT technology is to take on a more active defensive role.

- *Europe*: Europe is seeing a hiring boom in military industry big data roles. The European Defense Agency has made numerous recommendations about how the data is to be analyzed and thereafter fed to each combatant. The Europeans have included Modelling and Simulation (M&S) applications over the Cloud and the utilization of predictive analysis data in the development of M & S models.

- *India:* In an era where information is power, the Indian military is harnessing the potential of big data analytics to transform intelligence gathering. As the forces of developed countries, in the Indian Armed forces too, intelligence is at the heart of all defense planning and implementation. The major need of the hour is Big Data Strategy as well as recognition of its crucial importance from the top echelons to the front line soldier. Big data provides the Indian Army with advanced tools (analytics and algorithms) to reveal critical information. Handling of logistics data is a nightmare for the Indian Army [20]. Big data analytics is a game-changer in modern military intelligence, and the Indian Armed Forces are at the forefront of leveraging this technology. Through continuous innovation and strategic investment, the Indian military is poised to lead the way in leveraging big data analytics for national defense and security.

- *Ukraine:* Ukrainian military officials are evaluating whether to launch a cyber-operation to delete large quantities of data stored on a server run by a Russian company, located in Russia. Some of the data feeds into Russian weapons systems that operate within the armed conflict in Ukraine. But the proposed operation would also likely delete some civilian data that is stored on the same server. How would the existing rules apply to the data-deletion operation? Consider the efforts by Ukrainian officials to investigate and prosecute war crimes. Large quantities of data retrieved from a variety of sources such as satellites, social media, and crowdsourced information might be used to build a case against a particular perpetrator. Yet, private companies might own significant quantities of this data, creating challenges for prosecutors seeking to access it [21].

- *South Korea:* The Ministry of Defense is teaming up with the Ministry of Science, ICT, and Future Planning for a big data research project that will make use of the medical information of 600,000

South Korean military personnel. Currently, there are 19 military hospitals and some 1,200 medical detachments across the country, treating on average 2,500 soldiers daily. Regarding privacy concerns for the soldiers whose data will be utilized, the ministry emphasizes that protection of private information will be its foremost priority, and that all private information will be encrypted before being used for analysis [22]. South Korean military authorities have decided to add new food choices at military cafeterias based on a big data analysis of military food services. As illustrated in Figure 7.9, South Korean military uses big data for food service management [23].

Figure 7.9 South Korean military uses big data for food service management [23].

- *China:* China has wealth of data on what individuals are doing at a micro level. Pretty much everything about you is known or easily can be known by the government. There is an entire network, the Internet inside China's Great Firewall, designed to gather the information. Every picture posted, every comment made, every driving infraction or incident of rowdiness would go into a central database. Much of the data seems to come from companies like telecom providers and hotels. The approach is a far cry from what many Western governments still consider an appropriate balance between privacy and "national security." Technically, privacy is protected by the constitution and the law in China [24]. China is building up its arsenal of missiles, jets, and ships. Beijing's new missiles, aircraft carriers, and hypersonic glide vehicles threaten US military assets in China's littoral waters and beyond. Figure 7.10 shows ranges of Chinese land-based missiles [25]. Beijing developed at least two interrelated operational concepts aimed at dominating the information environment: system destruction warfare and multi-domain precision warfare. System destruction warfare targets vulnerable links between sensors and platforms. The system allows militaries to communicate, process battlefield information, and carry out strikes. Chinese strategists believe that the enemy will lose "the will and ability to resist." Multi-domain precision warfare aims to integrate AI and big data analysis with precision strikes to identify and target enemy weaknesses [26].

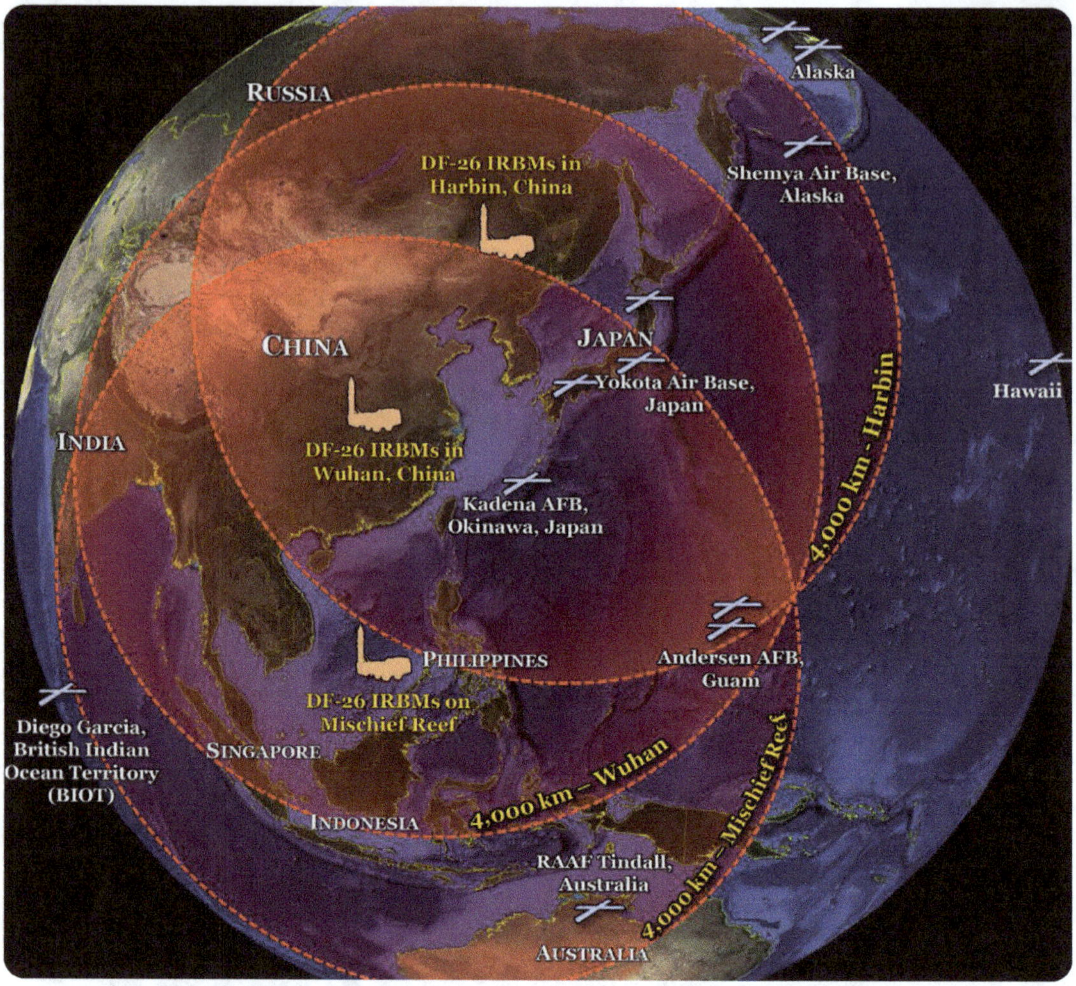

Figure 7.10 Ranges of Chinese land-based missiles [25].

7.7 BENEFITS

Big data is used in a variety of battlefield functions, such as targeted killing operations and intelligence collection and analysis. It is used to improve the procurement, transportation, and redeployment of personnel and material. Skillful deployment of big data analytics has the potential to confer a competitive edge amid complex and ever-evolving operational environments. Other benefits of big data in the military include the following:

- *Flexible Decision-making:* Data-driven decision-making enables military personnel to utilize real-time information to make decisions while personnel are in the field or on the base waiting for assignments. Big Data provides decision-makers with timely and accurate information for effective decision-making. Considering the rise of big data, decision-makers and commanders must grasp the potential of this data and its inherent utility on the battlefield. The abundance of available data can significantly augment intelligence analysis and bolster more knowledgeable and flexible decision-making processes during military operations. Army decision-making must remain widely distributed to maintain tactical flexibility.

- *Logistics:* Big Data will also play an important part in the fields of human resources and logistics. Big data aids in optimizing military logistics. Logistics is the backbone of any military; without it, the military is rendered inoperable. By analyzing historical data on supply chains, transportation routes, and maintenance records, military planners can identify bottlenecks, streamline operations, and improve resource utilization and logistics.

- *Democratization of Data:* This is the act of making the data easily accessible to those who need it. It is another internal push driving the utilization of big data in defense. However, more easily accessed data comes with its own set of challenges, primarily revolving around security concerns and system innovation.

- *Improved Decision-making*: Big data can help military leaders make more informed decisions by providing real-time insights and situational awareness.

- *Faster Response Time:* Big data can help the military respond quickly to threats by identifying them before they escalate.

- *Optimized Operations:* Big data can help the military optimize operations by analyzing data on logistics, resource allocation, and troop deployment. It can help the military understand the enemy by analyzing their communication, movement patterns, and social media activity. It can help the military develop more effective weapons by analyzing data to identify specific threats.

- *Improved Surveillance:* Big data can help the military develop sophisticated surveillance systems that can monitor in real-time.

7.8 CHALLENGES

While big data holds significant potential for revolutionizing defense intelligence, its exploration necessitates a nuanced consideration of both limitations and ethical considerations. It is crucial to address challenges related to data security, interoperability, and ethical considerations to ensure responsible and effective utilization of big data for military purposes. Other challenges of big data in military operations include the following [19]:

- *Ethical Concerns:* Ethical considerations arise regarding the use of data in military operations. Questions of proportionality, accountability, and the potential for biased decision-making require careful attention to ensure the responsible and ethical use of big data in warfare. The expanding influence of big data analytics on strategic decision-making within military operations necessitates a robust dedication to ethical considerations and methodological precision. Ethical concerns revolve around potential safety hazards, the responsible use of data, safeguarding privacy, and ensuring data protection. Successfully harnessing the transformative potential of big data within defense intelligence management requires a comprehensive approach that addresses ethical principles.

- *Expertise:* The precise understanding and effective utilization of data analytics require skilled professionals. Analysis needs expert analysts. Data-intensive fusion and analysis always require expert analysts to make sense of outputs. Even puzzles require expert analysts—to frame the puzzles in the first place, solve them, and then to make them relevant. For mysteries, data may offer valuable piecemeal insights, but expert analysts need to do even more heavy lifting to translate those insights into meaningful assessments for customers. Expertise is critical for inferring a target commander's intent.

- *Inefficiency:* The US military has staff that attempts to tackle processing their data. However, a human being can work at a human speed. This is a huge limitation and inefficiency. Between the joint headquarters and a battalion, there are six tiers of leadership at minimum. Before they can decide, a joint capability request will need to travel through every tier of leadership, which again, is at least six.

- *Data Security and Privacy Concern:* This remains the paramount concern as militaries collect and store vast amounts of sensitive information. Safeguarding this data against cyber threats and unauthorized access is crucial.

- *Interoperability:* Another challenge is the integration and interoperability of diverse data sources. Different military branches and agencies may use disparate systems, making it essential to establish data standards and frameworks for seamless data sharing and collaboration. In order for both industry and DoD officials to be successful in leveraging big data successfully, it is increasingly clear that open standards and interoperability will be key, especially with the push toward more cross-domain access to that data.

- *Information Overload:* The US military is struggling with information overload. Big data in the military comes from many sources and information overload is a very real problem. Increasingly, weapons systems depend on unfathomably large quantities of data to operate. Technologies that process and analyze large quantities of data, including artificial intelligence and machine learning, can exponentially increase military capabilities.

- *Bias:* The role of big data on the battlefield also raises important questions about privacy rights, discrimination, and bias

7.9 CONCLUSION

Big data refers to data that is too large or complex, grows, or changes at such a high velocity that traditional methods can no longer analyze it. It is emerging as a key component of military operations, both on and off the battlefield. On the battlefield, big data is being deployed, and has the potential to be deployed, for an astonishing array of purposes. Big data is also increasingly essential to military detention operations. It is now a key tool to investigate and prosecute those responsible for wartime atrocities. Big data therefore has the potential not only to revolutionize the tools armed forces use to fight, but to transform members of the armed forces themselves. More information about big data in military operations can be found in the books in [27-32] and the following related journals:

- *Military Review*
- *Journal of Military Learning*
- *Journal of Defense & Security Technologies*
- *The Cyber Defense Review*
- *Application of Big Data for National Security*

REFERENCES

[1] R. Delgado, "The challenges of bringing BYOD to the military," https://socpub.com/articles/the-challenges-of-bringing-byod-to-the-military-11272

[2] J. Moorthy et al., "Big data: Prospects and challenges," *The Journal for Decision Makers*, vol. 40, no. 1, 2015, pp. 74–96.

[3] "How does big data promise to transform the battlefield?" https://euro-sd.com/2024/06/articles/technology/38672/how-does-big-data-promise-to-transform-the-battlefield/#:~:text=By%20leveraging%20advanced%20analytics%20techniques, competitive%20 edge%20on%20the%20battlefield.

[4] M. N. O. Sadiku, C. M. M. Kotteti, and J. O. Sadiku, "Big data in the military," *International Journal of Trend in Research and Development,* vol. 11, no. 5, September-October 2024, pp. 58-65.

[5] M. Swan, "Philosophy of big data," *Proceeding of IEEE First International Conference on Big Data Computing Service and Applications*, 2015, pp. 468-477.

[6] "The complete overview of big data," https://intellipaat.com/blog/tutorial/hadoop-tutorial/big-data-overview/

[7] R. Allen, "Types of big data | Understanding & Interacting with key types (2024)," https://investguiding-com.custommapposter.com/article/types-of-big-data-understanding-amp-interacting-with-key-types

[8] M. N.O. Sadiku, M. Tembely, and S.M. Musa, "Big data: An introduction for engineers," *Journal of Scientific and Engineering Research,* vol. 3, no. 2, 2016, pp. 106-108.

[9] X. Wu et al., "Knowledge engineering with big data," *IEEE Intelligent Systems*, September/October 2015, pp.46-55.

[10] "The 42 V's of big data and data science," https://www.kdnuggets.com/2017/04/42-vs-big-data-data-science.html

[11] P. K. D. Pramanik, S. Pal, and M. Mukhopadhyay, "Healthcare big data: A comprehensive overview," in N. Bouchemal (ed.), *Intelligent Systems for Healthcare Management and Delivery*. IGI Global, chapter 4, 2019, pp. 72-100.

[12] J. Moorthy et al., "Big data: Prospects and challenges," *The Journal for Decision Makers*, vol. 40, no. 1, 2015, pp. 74–96. https://www.grandviewresearch.com/industry-analysis/industrial-wireless-sensor-networks-iwsn-market

[13] A. K. Tiwari, H. Chaudhary, and S. Yadav, "A review on big data and its security," *Proceedings of IEEE Sponsored 2nd International Conference on Innovations in Information Embedded and Communication Systems*, 2015.

[14] M. B. Hoy, "Big data: An introduction for librarians," *Medical Reference Services Quarterly*, vol. 33, no 3. 2014, pp. 320-326.

[15] M. Viceconti, P. Hunter, and R. Hose, "Big data, big knowledge: Big data for personalized healthcare," *IEEE Journal of Medical and Health Informatics*, vol. 19, no. 4, July 2015, pp. 1209-1215.

[16] "Big data | Effective collecting, analyzing, using data critical for success on the battlefield," November 2023, https://www.army.mil/article/270894/big data effective collecting analyzing using data critical for success on the battlefield#:~:text=United%20States%20Army-,Big%20Data%20%7C%20Effective%20Collecting%2C%20Analyzing%2C%20Using%20Data%20Critical,for%20Success%20on%20the%20Battlefield&text=Over%20the%20past%20few%20decades,and%20improve%20decision%2Dmaking%20processes.

[17] "The role of data analytics in defense strategies," March 2024, https://medium.com/@analyticsemergingindia/the-role-of-data-analytics-in-defense-strategies-7810ed837848

[18] "The role of big data in military operations: A game-changer in modern warfare," https://globmill.com/the-role-of-big-data-in-military-operations-a-game-changer-in-modern-warfare/

[19] S. Jang, "The role of data collection and big data in military war planning," November 2023, https://medium.com/ciss-al-big-data/data-is-constantly-generated-a-collected-in-the-modern-era-by-everything-3738f0d8ac10

[20] P. K. Chakravorty, "Big data: Implications for the Indian army," January 2020, https://bharatshakti.in/big-data-implications-for-the-indian-army/#:~:text=Big%20data%20provides%20the%20Indian,operations%2C%20training%20and%20other%20activities.

[21] L. A. Dickinson, "Lieber studies big data volume – Big data and armed conflict – Legal issues above and below the armed conflict threshold," January 2024, https://lieber.westpoint.edu/big-data-armed-conflict-legal-issues-above-below-armed-conflict-threshold/

[22] Korea Bizwire, "Military uses big data for disease control and prevention," December 2016, http://koreabizwire.com/military-uses-big-data-for-disease-control-and-prevention/71669

[23] Korea Bizwire, "S. Korean military uses big data for food service management," December 2019, http://koreabizwire.com/s-korean-military-uses-big-data-for-food-service-management/150341

[24] "Big Brother collecting big data — and in China, it's all for sale," January 2017, https://www.cbc.ca/news/world/china-data-for-sale-privacy-1.3927137

[25] "Big data for big wars: JEDI vs. China & Russia," August 2019, https://breakingdefense.com/2019/08/big-data-for-big-wars-jedi-vs-china-russia/

[26] S. Bresnick, "China bets big on military AI," https://cepa.org/article/china-bets-big-on-military-ai/

[27] M. N. O. Sadiku, U. C. Chukwu, and P. O. Adebo, *Big Data and Its Applications.* Moldova, Europe: Lambert Academic Publishing, 2024.

[28] L. A. Dickinson and E. W. Berg (eds.), *Big Data and Armed Conflict: Legal Issues Above and Below the Armed Conflict Threshold (The Lieber Studies Series).* Oxford University Press, 2024.

[29] K. Huggins (ed.), *Military Applications of Data Analytics.* Boca Raton, FL: CRC Press, 2023.

[30] G. Galdorisi and S. J. Tangredi (eds.), *AI at War: How Big Data, Artificial Intelligence, and Machine Learning Are Changing Naval Warfare.* Naval Institute Press, 2021.

[31] N. Lim, B. R. Orvis, and K. C. Hall, *Leveraging Big Data Analytics to Improve Military Recruiting.* RAND Corporation, 2019.

[32] E. Berman, J. H. Felter, and J. N. Shapiro, *Small Wars, Big Data: The Information Revolution in Modern Conflict.* Princeton University Press, 2018.

CHAPTER 8

BLOCKCHAIN IN THE MILITARY

"Blockchain facilitates democratic decision-making by enabling collective rule-setting and building in checks and balances."
– Lan van Wassenaer

8.1 INTRODUCTION

Contracts, transactions, and their records are critical, defining structures in our economic, legal, and political systems, but they have not being able to keep up with the world's digital transformation. Blockchain (BC) promises to solve this problem. It came as a solution to the longstanding user's trust problem. It is a technology that builds a trustworthy service in an untrustworthy environment. It refers to a highly secure and decentralized ledger system on which information can be stored but cannot be altered. It has evolved beyond cryptocurrencies to general purpose and can be used across an array of applications. It has proven to be another revolutionary technology that will impact many industries and transactions.

Blockchain (also known as "distributed ledger technology") is a peer-to-peer network that sits on top of the Internet. Blockchain technology is an innovation which is regarded as the center of Industry 4.0 revolution and it has become part of our lives. It is a system that stores data in a special way. Blockchain technology has some interesting properties, such as its decentralized nature, immutability, decentralization, transparency, and permissionless, that may be used to address pressing issues in many sectors. Although this technology finds its first application in the financial sector, it has become possible to use it in all sectors which can be integrated with technology today.

Blockchain technology is digitally signed and time-stamped data clusters that are published and linked together like a chain. It allows multiple users to publish at the same time through a secure algorithm in multiple cyber locations without the risk of data manipulation. It is one of the emerging technologies with the potential to transform society, especially the military. Blockchain is drawing the attention of researchers due to their potential applications beyond finance where it emerged. It is a revolutionary development that has transformed how military defense and operations are conducted. It is a promising technology that can have immense potential to provide decentralized trust, data security and integrity, traceability, transparency, visibility, and auditability across various areas in defense industries [1].

In this chapter, we explore how Blockchain could play a pivotal role in shaping the future of the military Army. The chapter begins with explaining what Blockchain is all about. It discusses military Blockchain and provides some of its applications. It covers military Blockchain around the world. It highlights the benefits and challenges of military Blockchain. The last section concludes with comments.

8.2 WHAT IS BLOCKCHAIN?

Blockchain, a type of distributed digital ledger technology (DLT), is a relatively new and exciting way of recording transactions in the digital age. It is a decentralized and distributed digital ledger technology that securely records and verifies transactions across multiple computers or nodes in a network. Basically, it is a chain of blocks in which each block contains a list of transactions. The blockchain technology was created as the foundational basis for Bitcoin – a digital currency in which secure peer-to-peer transactions occur over the Internet. It is expected that the spending on blockchain solutions worldwide would grow from 4.5 billion USD (2020) to an estimated value of 19 billion USD by 2024 [2].

Before Blockchain technology, people turned to gold or real estate when inflation hit its peak. Today, governments all over the globe have started opening up to Blockchain and crypto. By using blockchain, governments can reduce administrative costs, increase transparency, and improve service delivery. Blockchain is revolutionizing the digital world by bringing a new perspective to security, efficiency, and stability of systems and data. It is network of computers that is decentralized. Blockchain keeps track of distributed data and provides encrypted transaction tracking. It has attracted attention with its unique characteristics, such as irrevocability and security. It will be a part of our everyday life. Technology giants, such as IBM, Accenture, and Goldman Sachs are focusing on utilizing blockchain technology to enable cost-effective, secure, and transparent business operations [3,4].

Originally developed as the accounting method for the virtual currency Bitcoin, Blockchains are appearing in a variety of commercial applications today. Blockchain technology is a type of distributed digital ledger that uses encryption to make entries permanent and tamper-proof and can be programmed to record financial transactions. It is used for secure transfer of money, assets, and information via a computer network such as the Internet without requiring a third-party intermediary. It is now being adopted across financial and non-financial sectors. As a catalyst for change, the Blockchain technology is going to change the business world and financial matters in major ways.

The first Blockchain was conceived in 2008 by an anonymous person or group known as Satoshi Nakamoto, who published a white paper introducing the concept of a peer-to-peer electronic cash system he called Bitcoin [5,6]. Bitcoin and Ethereum are the first two mainstream Blockchains. Other modern Blockchains include Namecoin, Peercoin, Ether, and Litecoin. Figure 8.1 shows different components of Blockchain [7].

Figure 8.1 Different components of Blockchain [7].

Blockchain combines existing technologies such as distributed digital ledgers, encryption, immutable records management, asset tokenization and decentralized governance to capture and record information that participants in a network need to interact and transact. As illustrated in Figure 8.2, a complete blockchain incorporates all the following five elements [7]:

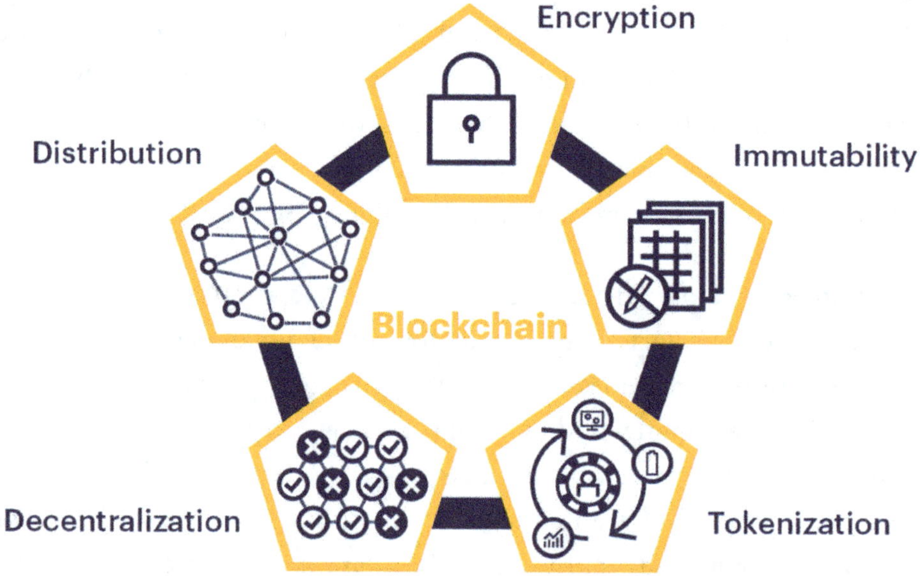

Figure 8.2 Five key elements of Blockchain [8].

- *Distribution:* Digital assets are distributed, not copied or transferred. A protocol establishes a set of rules in the form of distributed mathematical computations that ensures the integrity of the data exchanged among a large number of computing devises without going through a trusted third party. A centralized architecture presents several issues including a single point of failure and problems of scalability.

- *Encryption:* BC uses technologies such as public and private keys to record data securely and semi-anonymously. Completed transactions are cryptographically signed, time-stamped, and sequentially added to the ledger.

- *Immutability:* The Blockchain was designed so these transactions are immutable, i.e. they cannot be deleted. No entity can modify the transaction records. Thus, Blockchains are secure and meddle-free by design. Data can be distributed, but not copied.

- *Tokenization:* Value is exchanged in the form of tokens, which can represent a wide variety of asset types, including monetary assets, units of data or user identities.

- *Decentralization*: No single entity controls a majority of the nodes or dictates the rules. A consensus mechanism verifies and approves transactions, eliminating the need for a central intermediary to govern the network.

Bitcoin and its underlying blockchain technology increasingly impact all facets of society. Bitcoin's status as digital gold is merely the tip of this technology. Figure 8.3 shows Bitcoin [9]. Although blockchain technology will for all time be associated with Bitcoin due to their common genesis, it has broader applications. Cryptocurrency will increasingly become a factor in family law issues as well.

Figure 8.3 Bitcoin [9].

8.3 MILITARY BLOCKCHAIN

Blockchain has the potential to transform several industries, including the military. Integration of blockchain into military operations can significantly improve security, resource allocation, fraud reduction, and operational resilience. Data sharing through a blockchain can increase trust in detailed accounts, improve seamless communication, reduce data variation, and mitigate friction points when information transfer needs to be timely and actionable. Blockchain technologies can also support food safety and health care challenges on the battlefield, build health data sharing platforms for increased security and efficiency, track, and trace the food supply chain to prevent food related outbreaks.

Blockchain data immutability drastically improves the security of data being shared across all branches of the military. It could also help secure personal data confidentiality, 3D printing data, soldiers registers or smart contracts. The smart contracts can be used to track ammunition supply in a battlefield, verify aircraft components configuration, authenticate battlefield entities through registration, certify crew certificate, and audit battlefield operations.

8.4 APPLICATIONS OF MILITARY BLOCKCHAIN

In a secure environment, blockchain has potential for military application at each planning level and across all supply classes. Blockchain technology is being used in the military for a variety of purposes such as information security, authentication, data integrity, resilience, operations management, supply chain management, logistics. etc. The specific applications include the following [10-13]:

1. *Managing Supply Chain:* The Department of Defense's (DOD) supply chain network is one of the largest and most complex in the world, with thousands of supply chain partners across the globe. The defense supply chain is vital to national security. Secure supply chain management ensures timely availability of equipment, materials, and information for military response to threats and emergencies. Blockchain has the potential to significantly transform DOD supply chain networks. It may help military organizations ensure their supply chains are robust and secure. Adding blockchain technology to defense supply chain management systems will improve military security, efficiency, and transparency. Defense supply chain management application security is a global military priority. Blockchain's decentralized, immutable, and trust-based nature can help to improve the security and transparency of defense supply chains. Its decentralized, immutable, and trust-based characteristics can improve defense supply chain security and transparency. For example, blockchain can be used to record every transaction in a weapon's supply chain, creating a transparent history that can be audited by authorized stakeholders. Blockchain may offer the Army a solution it needs to secure the "digital thread" integral to the additive manufacturing supply chain. However, integrating blockchain into military operations requires a holistic approach that considers technology, regulatory compliance, data privacy, training, and maintenance.

2. *Military Logistics:* The pentagon has the world's largest logistics problem. In the context of data sharing within military logistics, blockchain technology is emerging as a valuable tool, due to its decentralized nature and data security capabilities. The development of blockchain technology offers increased data confidence and data availability that can help shape future military logistics and planning. Blockchain technology has significant potential value for complex logistics applications in commercial, public, and military environments. It eliminates data integrity issues and prevents data tampering by providing a single source that all stakeholders can access simultaneously. This makes data sharing much simpler while reducing human errors and time lags occurring when utilizing traditional methods.

3. *Military Defense*: In an increasingly digital world, the defense of military systems and networks is of utmost importance. Blockchain technology can offer military defense forces a wide range of benefits, particularly in the areas of military logistics and supply chain management. Blockchain can help to protect military communications from cyber threats by making them more secure,

transparent, and tamper-proof. Its cryptographic security capabilities can improve military communications by adding an extra layer of authentication for sensitive messages and data stored online. Blockchain's cryptographic hashing and consensus algorithms make it difficult to alter or delete recorded data without the agreement of the network participants. This creates a permanent audit trail that helps to protect critical military information. Moreover, Blockchain technology has the potential to enhance end-to-end visibility into military operations so that planners can make more informed decisions.

4. *Data Management:* Data management is another area of application. It is paramount to federal communications. One of the most important components of success in military defense is data management. Traditional methods of storing and managing data are susceptible to hacking and fraud. Blockchain technology provides a potential solution since it is much more secure. By improving data integrity through the use of distributed ledgers and automation, it can reduce manual tasks significantly to better manage complex networks. Blockchain can be used to register every entity on the battlefield, including information about vehicles, aircraft, weapons, and warfighters. With blockchain, data is stored in a distributed ledger that is incredibly difficult to tamper with. This makes it an ideal solution for storing sensitive information such as military personnel data including basic information, career path, missions undertaken, and rewards received.

5. *Asset Tracking:* This is another area where blockchain technology could have a significant impact. In the military, it is important to know the location of all assets, including weapons, vehicles, and communications equipment. Blockchain-based asset tracking would allow for real-time tracking of assets and could help to prevent their loss or theft. With blockchain, each asset can be given a unique identifier that cannot be altered. This would allow the military to keep track of its assets with a high degree of accuracy. Blockchain could also be used to track the movement of assets throughout the supply chain. This would provide greater transparency and visibility into the supply chain, which would help to prevent theft and other forms of loss. Figure 8.4 shows a vehicle park of US Army [14].

Figure 8.4 A vehicle park of the US Army [14].

6. *Cybersecurity:* Another application is cybersecurity. Cybersecurity is a critical concern for the military, as hackers can potentially gain access to classified information or disrupt critical infrastructure. In today's hyper-connected world, military communication systems face relentless threats from cyberattacks, espionage, and data breaches. Blockchain technology could be used to create a secure, decentralized network that would be resistant to cyber-attacks. Blockchain technology provides greater protection against cyber attacks and threats. The military could implement blockchain to secure data from unauthorized access or alteration. Incorporating blockchain into an effective cybersecurity strategy would significantly strengthen the defense against malicious adversaries in times of crisis.

7. *Registration Process*: The blockchain-based registration process for defense applications can help to register every battle-field entity. This includes information from various sensors, drones, combat equipment, vehicles, aircraft, smart weapons, and warfighters along with their role in a military operation. Blockchain employs smart contracts for the registration process to ensure that only registered members can read the data stored in the ledger. It also defines the access rights for every battlefield warfighter as it depends on their position in the military hierarchy.

8.5 MILITARY BLOCKCHAIN AROUND THE WORLD

Governments across the world are showing interest in blockchain technology. They are starting to embrace blockchain as a secure way to store and manage large amounts of vital data while simultaneously improving overall operational effectiveness. Many governments have implemented blockchain applications within their military forces to improve the organization's data security systems. It is now for sure that military organizations around the world would want to ensure that they stay ahead of others. We consider how some nations employ Blockchain in their military organizations.

- *United States:* The United States Marine Corps adopted an extensive research project that provides its members with an encrypted platform powered by blockchain technology with which they can securely communicate and exchange data with both allied nations and domestic stakeholders. The US Defense Advanced Research Projects Agency (DARPA) is planning to implement blockchain technology in defense and military applications. The US military is already using blockchain in some of its applications [15].

- *China:* Chinese soldiers could be rewarded in cryptocurrency tokens for good performance if the country's armed forces embrace Blockchain technology. Applying the technology to managing the People's Liberation Army (PLA) would drive innovation. Government subsidies received by leading players in China's semiconductor industry have increased significantly. Beijing has doubled efforts to boost technological self-sufficiency amid growing tensions with Washington [16]. The Chinese military could adopt blockchain technology to manage personnel data, boost training and mission performances, and provide soldiers with earned tokens that could be used to collect rewards.

- *India:* India boasts the second-largest and fourth-strongest army globally. Its defense capabilities are evolving with the integration of emerging technologies like artificial intelligence, big data, and blockchain. The journey towards integrating blockchain into the Indian Army has just begun, promising a future where data manipulation is thwarted, decision-making is fortified, and the defense sector stands resilient in the face of evolving threats. The Defense Minister has emphasized the changing character of warfare and the need to prepare for contingencies while repelling threats from multiple sources. India wants to stay ahead of the curve and ensure that its military remains at the forefront of technological innovation [17].

- *Ukraine:* Crypto was supposed to be Ukraine's launchpad into the future. Instead it is proving to be a necessary lifeline in a country ravaged by war. Since Russia's invasion, Ukraine has raised more than $56 million in donations spread across assets such as bitcoin, ether, polkadot, solana, dogecoin, tether, etc. These funds have gone to help humanitarian agencies distributing aid in the country, procure necessary supplies for soldiers such as food, uniforms, and bullet-proof vests. Digital assets and blockchain technology are meant to help revitalize the Ukrainian economy and bring all government processes online. Blockchain initiatives have been taken to legalize digital assets in the country and make Ukraine one of the most crypto-friendly countries in the world. However, all of those plans went out the window with the Russian invasion. Figure 8.5 shows soldiers in the Ukrainian army [18].

Figure 8.5 Soldiers in the Ukrainian army [18].

- *South Korea:* South Korea's Defense Acquisition Program Administration (DAPA) plans to improve the reliability of data in the arms industry by using Blockchain. The agency plans to use Blockchain for secure sharing of data about military acquisition projects. It is seeking to use a Blockchain platform to enable the secure sharing of data about military acquisition projects between relevant governmental organizations and to eliminate data forgery. DAPA also intends to reduce the paperwork associated with applying for state procurement projects, as well as unify its documentation [19].

- *Nigeria:* To address the challenge of certificate forgery and fraudulent issuance in Nigeria, an agency has unveiled an initiative that will use blockchain technology to authenticate National Youth Service Corps (NYSC) certificates. It pioneers the development of an advanced Blockchain-based certificate authentication system. The revolutionary approach is designed to ensure the integrity and validity of NYSC certificates, thereby enhancing transparency and trust in the certification process. Figure 8.6 shows some graduates serving the NYSC [20].

Figure 8.6 Some graduates serving the NYSC [20].

8.6 BENEFITS

Blockchain is a promising and emerging technology that can have immense potential to provide decentralized trust, data security and integrity, traceability, transparency, visibility, and auditability across various areas in the defense industries. It is revolutionizing the military industry by providing secure, efficient data sharing, and storage solutions. It has the potential to solve major problems in military logistics, such as tracking supplies, managing personnel records, and ensuring data security. Other benefits of blockchain in the military include the following:

- *Cost Reduction:* Blockchain can also save money by eliminating unnecessary middlemen who add little, if any, value. Reducing or eliminating manual processes and reducing fraud can all reduce operational costs and further increase efficiency. Cost reductions are anticipated in regards to information lags, duplication, personnel, movement times, storage, and inventory losses.

- *Improve Operational Performance:* Most defense platforms and systems are staggeringly complex. These assets are very mobile (globally) and require collaboration among a number of entities to keep them operational and mission-ready. Technology can help operators, manufacturers and suppliers "harden" the supply chain and improve operational performance through the entire life cycle, from raw material to retired asset.

- *Decentralization:* At its core, blockchain's decentralized architecture eliminates the most significant vulnerability of traditional systems: the single point of failure. In contrast to centralized systems, blockchain operates on a decentralized network, where data is not stored in a single location but spread across multiple nodes. This means that even if one node is tampered with or compromised, the integrity of the overall system remains intact.

- *Immutability:* One of blockchain's most compelling features is its immutability. Blockchain's inherent immutability (i.e. once data is recorded, it cannot be altered) ensures that every message sent through military channels remains untampered. This feature is critical in high-stakes defense scenarios, where the integrity of communication can be a matter of life and death.

- *Transparency:* Military operations require precise coordination, and blockchain provides real-time visibility into the communication process. By integrating blockchain, defense forces can monitor every stage of message transmission, ensuring it has not been intercepted or altered. This transparency builds trust in the communication system, allowing faster and more confident decision-making during critical missions.

- *Advanced Encryption:* Blockchain technology uses advanced cryptographic techniques to secure data, making unauthorized access nearly impossible. Unlike traditional encryption methods, which can be cracked over time, Blockchain continually reinforces security through its distributed ledger, securing sensitive defense information.

- *Communication Security:* In an age where secure communication is paramount, blockchain technology is emerging as a game-changer. The decentralized approach makes it nearly impossible for attackers to infiltrate the network, offering unprecedented resilience and security for critical communications. Once data is recorded, it becomes a permanent, tamper-proof entry in the blockchain. For sectors like military communications or high-stakes corporate exchanges, this ensures that records remain intact, verifiable, and immune to tampering. By integrating smart contracts, blockchain automates security protocols, ensuring that communication rules are enforced without relying on intermediaries. Figure 8.7 depicts Blockchain-based communication in military network [21].

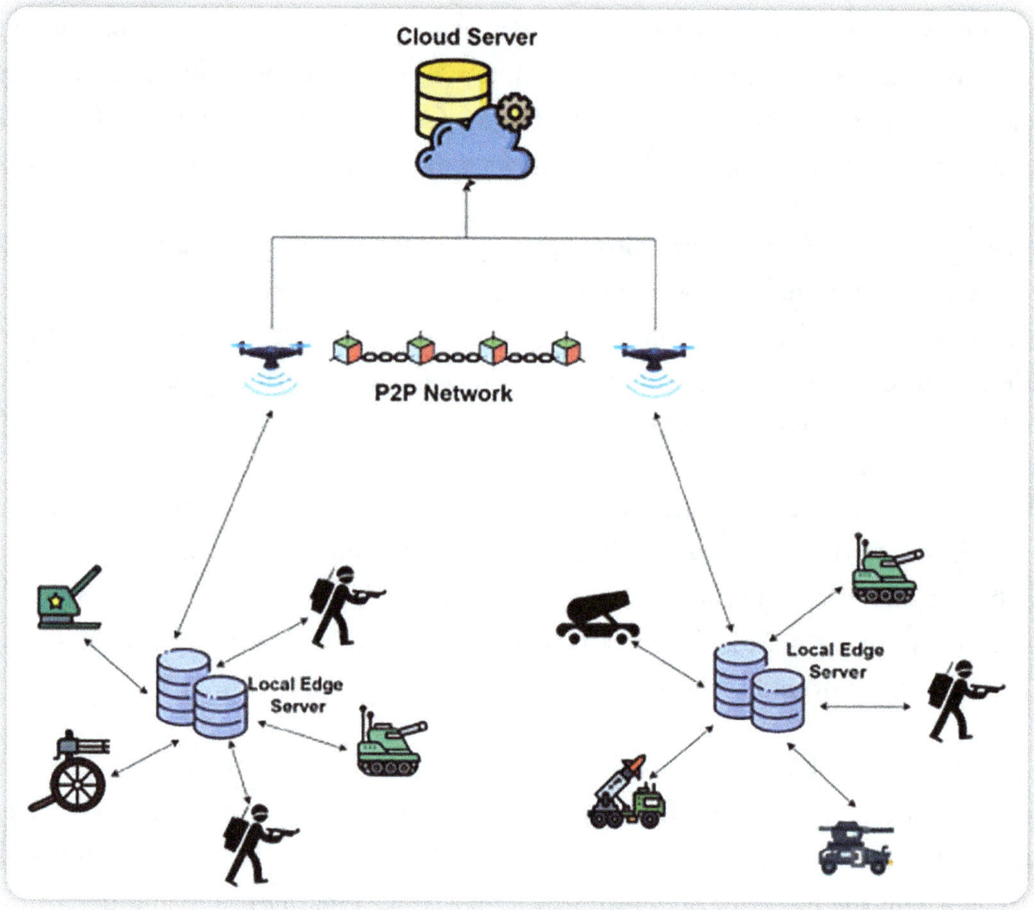

Figure 8.7 Blockchain-based communication in military network [21].

8.7 CHALLENGES

New technologies such as Blockchain create new challenges and opportunities. Challenges facing Blockchain may require further research on areas such as interoperability, network infrastructure, and a thorough analysis on its regulation. Other challenges of blockchain in the military include the following [10]:

- *Threats to Cybersecurity:* Blockchain integration into defense supply chain management can improve security, but it also shows the rising cybersecurity dangers. As supply chain applications become increasingly networked and data-intensive, sophisticated hackers target them.
- *Scalability:* To ensure successful implementation and long-term sustainability, defense supply chains must balance transparency and security with their complexity and scale. This requires careful planning, scalable blockchain infrastructure, and effective governance frameworks
- *Supply Chain Resilience:* Defense supply chains are vulnerable to natural disasters and geopolitical crises. Real-time visibility into commodities and asset movement on blockchain's decentralized ledger allows proactive risk assessment and mitigation, improving supply chain resilience.

- *Complexity:* The process of data integration in military logistics can be extremely complex and time-consuming, making it difficult for Blockchain ideation to align with the standards of data sharing within the sector. Military organizations need to develop an effective data integration process that is tailored to their individual needs.

- *Cost-effectiveness:* A major concern is the cost efficiency of incorporating data from existing systems into Blockchain-based ones, while also maintaining a fluid process. Military logisticians must pay attention to cost efficiency when implementing a Blockchain solution for military logistics, in order to ensure optimal performance without breaking the bank.

- *Security:* Smart contracts on blockchains can be vulnerable to security breaches. Decentralization can also pose a security risk, as it could jeopardize the information stored on the blockchain.

- *Cultural Shift*: Adopting blockchain technology may require a cultural shift in the military's approach to data management and security. Military personnel need adequate training to use blockchain technology effectively and securely.

8.8 CONCLUSION

The use of blockchain technology is gaining traction in all industries. It is being used across many industries for many purposes. Organizations around the world are beginning to recognize the value that blockchain technology can bring to their operations. Blockchain is set to transform military communications, offering unprecedented security, transparency, and resilience against cyber threats. As we navigate the complexities of modern warfare, Blockchain technology emerges as a strategic ally, offering unique advantages over conventional cyber defense approaches. However, the nascency of the technology, coupled with relative lack of knowledge of DoD concerning it, has delayed its adoption and integration by the DoD. The future possibilities for blockchain technology in militarized data sharing and logistics operations are exciting. More information about blockchain in military operations can be found in the books in [22-25] and the following related journals:

- *Military Review*
- *Journal of Military Learning*
- *Journal of Defence & Security Technologies*
- *The Cyber Defense Review*

REFERENCES

[1] M. N. O. Sadiku, C. M. M. Kotteti, and J. O. Sadiku, "Blockchain in the military," *Journal of Scientific and Engineering Research*, vol. 11, no. 9, pp. 106-114.

[2] C. M. M. Kotteti and M. N. O. Sadiku, "Blockchain technology," *International Journal of Trend in Research and Development*, vol. 10, no. 3, May-June 2023, pp. 274-276.

[3] A. Guadamuz and C. Marsden, "Blockchains and Bitcoin: Regulatory responses to cryptocurrencies," *Peer-reviewed Journal on the Internet*, vol. 20, no. 12, Dec. 2015.

[4] M. N. O. Sadiku, Y. Wang, S. Cui, and S. M. Musa, "A primer on Blockchain," *International Journal of Advances in Scientific Research and Engineering*, vol. 4, no. 2, February 2018, pp. 40-44.

[5] "Blockchain," *Wikipedia,* the free encyclopedia
https://en.wikipedia.org/wiki/Blockchain

[6] S. Nakamoto, "Bitcoin: A peer-to-peer electronic cash system,"
https://bitcoin.org/bitcoin.pdf

[7] "The beginning of a new era in technology: Blockchain traceability,"
https://www.visiott.com/blog/blockchain-traceability/#:~:text=The%20Beginning% 20of%20a%20New,money%20without%20a%20central%20bank.

[8] "The CIO's guide to Blockchain,"
https://www.gartner.com/smarterwithgartner/the-cios-guide-to-blockchain#:~:text=True%20blockchain%20has%20five%20elements,%2C%20immutability%2C%20tokenization%20and%20decentralization.

[9] "Helping the military adopt Blockchain technology,"
https://www.pioneeringminds.com/helping-military-adopt-blockchain-technology/

[10] K. Sahu and R. Kumar, "Blockchain for managing the supply chain in military operations," December 2023, https://www.bcs.org/articles-opinion-and-research/blockchain-for-managing-the-supply-chain-in-military-operations/

[11] M. T. Simerly and D. J. Keenaghan, "Blockchain for military logistics," November 2019,
https://www.army.mil/article/227943/blockchain for military logistics

[12] "Blockchain for military logistics: What you should know," April 2023,
https://nstxl.org/blockchain-for-military-logistics/#:~:text=April%204%2C%202023,records %2C%20and%20ensuring%20data%20security.

[13] H. H. Ahmad et al., "Blockchain for aerospace and defense: Opportunities and open research challenges," *Computer & Industrial Engineering*, vol. 151, January 2021.

[14] E. Broitman, "The Pentagon has the world's largest logistics problem. Blockchain can help," October 2017, https://www.defenseone.com/ideas/2017/10/pentagon-has-worlds-largest-logistics-problem-blockchain-can-help/141500/

[15] "How Blockchain solidifies defense and military applications," January 2018, https://blockchain.oodles.io/blog/blockchain-military-applications/

[16] "Reward Chinese soldiers in cryptocurrency, military mouthpiece says," https://www.scmp.com/news/china/military/article/3037592/reward-chinese-soldiers-cryptocurrency-military-mouthpiece-says

[17] P. Jain, "Unleashing the power of Blockchain: Transforming the Indian military landscape," November 2023, https://www.linkedin.com/pulse/unleashing-power-blockchain-transforming-indian-military-pratik-jain-ryo8f

[18] S. Ehrlich, "Crypto interrupted: How the Russian invasion dramatically changed Ukraine's Blockchain strategy to focus on the war," March 2022, https://www.forbes.com/sites/stevenehrlich/2022/03/05/crypto-interrupted-how-the-russian-invasion-dramatically-changed-ukraines-blockchain-strategy-to-focus-on-the-war/

[19] H. Partz, "South Korea State Defense Arm DAPA to build blockchain platform for military acquisition," April 2019, https://cointelegraph.com/news/south-korea-state-defense-arm-dapa-to-build-blockchain-platform-for-military-acquisition

[20] M. Akuchie, "Why issuing NYSC certificates through blockchain technology is not enough to drive adoption," August 2023, https://technext24.com/2023/08/23/blockchain-nysc-certificate-nigeria/

[21] M. Golam, J. Lee, and D. Kim, "A UAV-assisted blockchain based secure device-to-device communication in Internet of military things," October 2020, https://www.researchgate.net/publication/347805850

[22] M. N. O. Sadiku, *Blockchain Technology and Its Applications.* Moldova, Europe: Lambert Academic Publishing, 2023.

[23] U. S. Military, Department of Defense (DoD), T. Doskey, *Improve Acquisitions, Procurement, and Supply Chain, Analysis of Private Sector Programs, Purchase Card Pilot Program.* Amazon Digital Services LLC, 2019.

[24] U. S. Military, Department of Defense (DoD), N. Barnas, *Blockchains in National Defense: Trustworthy Systems in a Trustless World - The Evolving Cyber Threat, Air Force Should Research and Develop Blockchain Technology to Reduce Probability of Compromise.* Independently Published, 2019.

[25] V. Adams, *Potential Uses of Blockchain by the U.S. Department of Defense.* Value Technology Foundation, 2020.

CHAPTER 9

CYBERSECURITY IN THE MILITARY

"Cybersecurity is a subject that requires logic, knowledge, thought, and commitment. It can be applied or research based. It is a true leveler for all to enter, be successful and lead the future of cybersecurity."
– Ian R. McAndrew

9.1 INTRODUCTION

Access to the Internet is essential for global security and prosperity. The technological advancements of the Internet have enabled information to be more accessible to people all over the world. Because everything is networked and interconnected, a breach at any point could enable professional hackers to work their way up the food chain to more important targets, from identity theft to your bank account.

Being in the digital era, a cyber-terrorist assault is a possibility. Such a attack poses a threat to the peace and security of any nation. As technology advances, cyber threats against military organizations have grown and become sophisticated. These attacks can compromise sensitive information, disrupt military operations, and target critical infrastructure. Hackers and cybercriminals launch new, sophisticated computer viruses, malware, and scams daily that threaten the sensitive data our society relies on. Criminal and political cyberattacks have become more prominent in the media in recent years. Awareness is growing in policy circles, and many nations have developed their own national cybersecurity and defense strategies [1].

Cybersecurity refers to a set of behaviors, methods, and technologies aimed at protecting systems, networks, data, and computers from harm, attacks, and illegal access. Like any other institution, the military relies on cybersecurity to keep itself safe. Military forces have developed cyber capabilities to aid them in combat and to secure their networks from foes in times of peace. Military cybersecurity is critical in protecting both their troops and the citizens. The cybersecurity issue has taken the significant role and emerged as a planning factor almost in every public or private institution [2].

The primary responsibility of the military is to protect all citizens, but it cannot do that if it cannot protect itself. The military essentially protects our interests both at home and abroad. The widespread adoption of digital technologies has increased the likelihood of cyberattacks on the military. The military is making cybersecurity an operational priority ensuring resources support training, technology, and policies. The goal of the military's cybersecurity strategy is to mitigate any potential hazards to military operations. Our social security numbers and our nation's top secrets must be safeguarded and kept safe from any adversaries who could attempt to take advantage of them [3]. Our adversaries China, Russia, Iran, and North Korea are increasingly taking malicious cyber activities in the gray zone. China is using cyber espionage for military and economic advantages. Russia is conducting cyber espionage that has the potential to disrupt critical infrastructure. North Korea has hacked financial networks and cryptocurrency [4]. Iran has demonstrated the capability and intent to strike in its region and against the United States in cyberspace.

This chapter introduces readers to cybersecurity in the military domain. It begins with providing an overview of cybersecurity. It discusses military cybersecurity and how to protect the military. It highlights the benefits and challenges of cybersecurity in the military. The last section concludes with comments.

9.2 OVERFVIEW ON CYBERSECURITY

Cybersecurity refers to a set of technologies and practices designed to protect networks and information from damage or unauthorized access. It is vital because governments, companies, and military organizations collect, process, and store a lot of data. As shown in Figure 9.1, cybersecurity involves multiple issues related to people, process, and technology [5].

Figure 9.1 Cybersecurity involves multiple issues related to people, process, and technology [5].

A typical cyber attack is an attempt by adversaries or cybercriminals to gain access to and modify their target's computer system or network. Cyber attacks are becoming more frequent, sophisticated, dangerous, and destructive. They are threatening the operation of businesses, banks, companies, and government networks. They vary from illegal crime of individual citizen (hacking) to actions of groups (terrorists) [6].

The cybersecurity is a dynamic, interdisciplinary field involving information systems, computer science, and criminology. The security objectives have been availability, authentication, confidentiality, nonrepudiation, and integrity. A security incident is an act that threatens the confidentiality, integrity, or availability of information assets and systems [7].

- *Availability*: This refers to availability of information and ensuring that authorized parties can access the information when needed. Attacks targeting availability of service generally leads to denial of service.

- *Authenticity*: This ensures that the identity of an individual user or system is the identity claimed. This usually involves using username and password to validate the identity of the user. It may also take the form of what you have such as a driver's license, an RSA token, or a smart card.

- *Integrity*: Data integrity means information is authentic and complete. This assures that data, devices, and processes are free from tampering. Data should be free from injection, deletion, or corruption. When integrity is targeted, nonrepudiation is also affected.

- *Confidentiality*: Confidentiality ensures that measures are taken to prevent sensitive information from reaching the wrong persons. Data secrecy is important especially for privacy-sensitive data such as user personal information and meter readings.

- *Nonrepudiation*: This is an assurance of the responsibility to an action. The source should not be able to deny having sent a message, while the destination should not deny having received it. This security objective is essential for accountability and liability.

Everybody is at risk for a cyber-attack. Cyber-attacks vary from illegal crime of individual citizen (hacking) to actions of groups (terrorists). The following are typical examples of cyber-attacks or threats [8]:

- *Malware*: This is a malicious software or code that includes traditional computer viruses, computer worms, and Trojan horse programs. Malware can infiltrate your network through the Internet, downloads, attachments, email, social media, and other platforms. Spyware is a type of malware that collects information without the victim's knowledge.

- *Phishing*: Criminals trick victims into handing over their personal information such as online passwords, social security number, and credit card numbers.

- *Denial-of-Service Attacks*: These are designed to make a network resource unavailable to its intended users. These can prevent the user from accessing email, websites, online accounts or other services.
- *Social Engineering Attacks*: A cyber-criminal attempts to trick users to disclose sensitive information. A social engineer aims to convince a user through impersonation to disclose secrets such as passwords, card numbers, or social security number.
- *Man-In-the-Middle Attack*: This is a cyber-attack where a malicious attacker secretly inserts him/herself into a conversation between two parties who believe they are directly communicating with each other. A common example of man-in-the-middle attacks is eavesdropping. The goal of such an attack is to steal personal information.

These and other cyber-attacks are shown in Figure 9.2 [9].

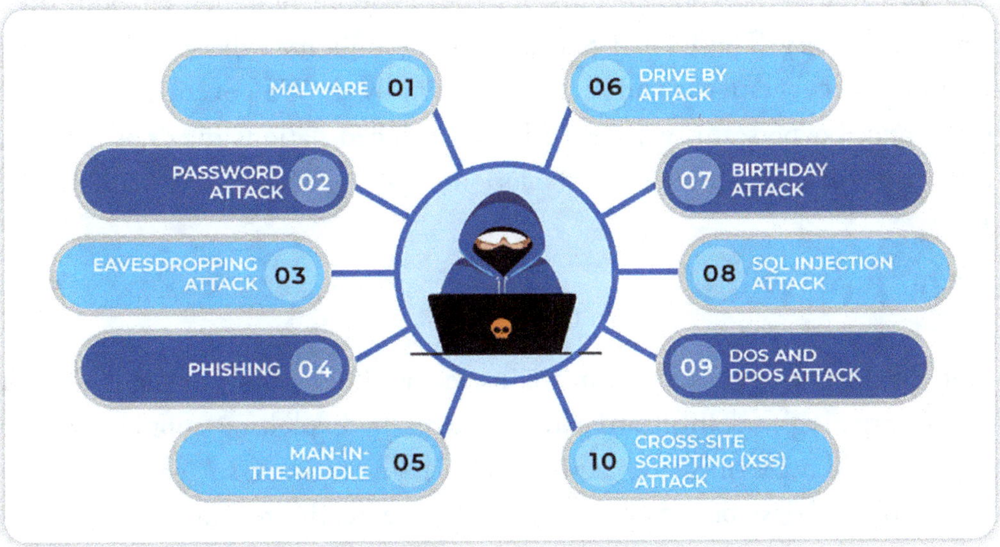

Figure 9.2 Common types of cyber attacks [9].

Cybersecurity involves reducing the risk of cyber-attacks. Cyber risks should be managed proactively by the management. Cybersecurity technologies such as firewalls are widely available [10]. Cybersecurity is the joint responsibility of all relevant stakeholders including government, business, infrastructure owners, and users. Cybersecurity experts have shown that passwords are highly vulnerable to cyber threats, compromising personal data, credit card records, and even social security numbers. Governments and international organizations play a key role in cybersecurity issues. Securing the cyberspace is of high priority to the US Department of Homeland Security (DHS). Vendors that offer mobile security solutions include Zimperium, MobileIron Skycure, Lookout, and Wandera.

9.3 MILITARY CYBERSECURITY

The military has developed cyber capabilities to help them in fighting in the battlefields and to defend their systems from enemies during peace time. They can now implement offensive actions to cyber-threats and hackers. Military cybersecurity operations are shifting to a digital battlefield, where tools and technology work to save lives and increase efficiency. Joint operations involve coordinating naval, air, and ground forces.

Military cybersecurity refers to the practice of protecting military organizations, assets, and operations from cyber threats and attacks. It aims to ensure the confidentiality, integrity, and availability of critical information, as well as to maintain the functionality and operational readiness of military systems. It involves measures such as network security, encryption, access controls, incident response, and threat intelligence to mitigate cyber risks and maintain national security [11].

Military organizations around the world are pursuing the development of new enabling, next-generation technologies for cybersecurity. The United States leads the military cybersecurity market due to the increase in the number of cyberattacks and high investment by the US government in the research and development of advanced cybersecurity systems. Less developed nations may not have the budgets to procure and implement advanced cybersecurity solutions.

9.4 PROTECTING THE MILITARY

Cybersecurity and cyberterrorism have become mainstream talking points in the United States. Today, our adversaries are targeting and infiltrating our systems by exploiting supply chain and zero-day vulnerabilities. The United States must also actively deter rogue regimes, such as North Korea and Iran, which are unpredictable and destabilizing presences in their respective regions. To manage cybersecurity for military systems/networks, the Army and the Department of Defense (DoD) need appropriate policies to foster system designs that are robust and resilient to cyber attacks. The US government, private industry, academia, and international partners must develop proactive defense strategies and contest our adversaries' campaigns and objectives. Such strategies are provided as follows.

- *Cybersecurity Readiness:* When it comes to national defense, readiness to execute missions matters a lot. The Army campaign is designed to increase readiness through improving awareness of cyber threats and incidents. The Coast Guard attack in 2019 was not a drill or hypothetical scenario. It happened and demonstrated what can occur in the absence of operational readiness. The incident illustrates the urgent need for military agencies to establish cybersecurity readiness for their systems. The operational readiness of industrial control systems (ICS) infrastructures enable the availability of power, water, and other functions at military bases. Cybersecurity readiness and active defenses should be part of military training [12]. Active defenses refer to the use of limited action to fend off an attack. In digital security terms, it means you go on the offensive to identify possible threats to your network and move first to neutralize the threat before it can cause any damage.

- *AI/ML Tools:* The cyberspace is so vast, so complex, so constantly changing that only artificial intelligence can keep up. Advances in artificial intelligence and machine learning will likely add to cybersecurity challenges in the future. A team of cyber expects have developed machine learning tools and tactics, techniques, and procedures for an analysis of DoD network traffic data [13]. Soldiers deployed on the battlefield will be shielded from cyberattacks without human involvement. Artificial intelligence can protect soldiers' tactical networks and communications from cyberattacks. Artificial intelligence and robotics are making their way into operations, from carrying tools to charting battle strategies. Machine learning can automatically detect known cyber vulnerabilities, spot previously unknown malware, and respond to a cyberattack.

- *Automation:* This creates a strategic imperative for automation: software programs that can detect vulnerabilities. But cybersecurity automation is yet to come of age. Computers or autonomous systems are going to replace humans.

- *Zero Trust Architecture:* The adoption of "zero trust" principles within the DoD is important. Industry and government experts recently decided to employ a zero trust architecture in all military operations. A zero trust architecture ensures that only the right people access the proper data by using continual verification throughout the system. It requires frequent authentication, authorization, and validation of users. Cloud technology is a major player in zero trust architecture [14].

- *Recruiting Veterans:* The cybersecurity industry is undergoing a severe talent shortage, putting our communities, economies, and nation at risk. Our nation needs more cybersecurity professionals at all levels, in every organization, and in every region. There are over 700,000 open cybersecurity jobs in the US alone. The demand for cybersecurity experts is growing faster than the current US job market, making cybersecurity one of the most highly sought-after careers in the nation. To help veterans enter this fast-growing, rewarding career, CISA provides Cybersecurity Training and Education for Veterans [15]. These men and women are some of the highest-skilled, best-trained, hardest-working people in the nation. Cybersecurity will be a great way to continue serving the country since veterans understand the value our society places on national security. It is widely recognized that veterans bring the proper ethos to their work. Top companies are actively looking for talented veterans and recruiting them for key cybersecurity jobs. As a veteran, you have a unique skillset that virtually any private organization can benefit from. Cybersecurity offers veterans a natural career progression. Figure 9.3 shows some of the character traits of valued veterans in cybersecurity [16]. These cybersecurity skills are focused on detecting, defending, responding to, and preventing cyber-attacks that may harm military systems, networks, and operations.

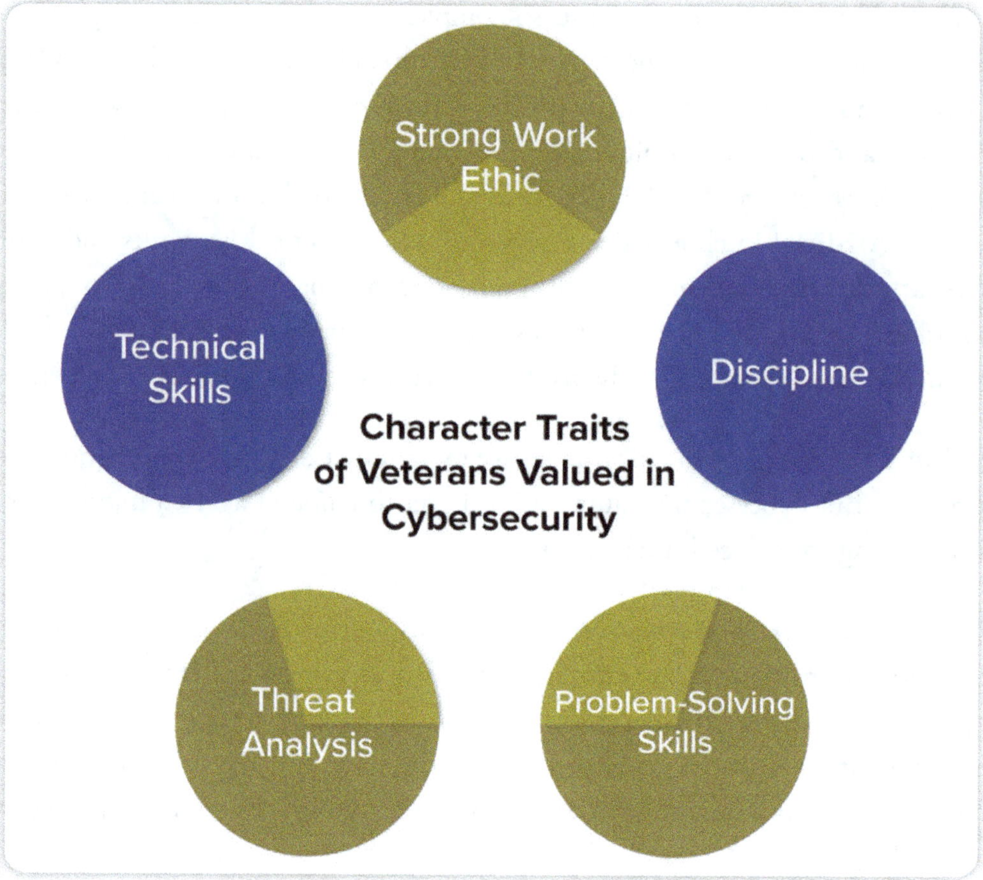

Figure 9.3 Some of the character traits of valued veterans in cybersecurity [16].

- *Cyber War:* The US Army is recruiting smart young soldiers to wage cyber war. Figure 9.4 shows some cyber warfare operators [17]. The overall mission of an Air Force Cyber Warfare operator is to create effects designed to deny, destroy, disrupt, degrade, or manipulate adversary capabilities. More than twenty years after the September 11, 2001 attacks, the great threat to the nation has evolved into cyberspace. Cyberspace is a global domain which consists of the interdependent networks of information technology infrastructures including the Internet, telecommunications networks, computer systems, satellite systems, and embedded processors and controllers. The entire world, especially the military, is firmly entrenched in cyberspace. Electric power grids, water treatment facilities, hospitals, traffic control, aviation, railroads, etc. are all part of the cyber domain. For the past two decades, US forces have continued to dominate cyberspace and the electromagnetic spectrum while conducting counterinsurgency operations in Afghanistan and Iraq against enemies and adversaries who lack the ability to challenge our technological superiority. By operating in cyberspace, US adversaries can cause damage and they are targeting US economies, critical infrastructure, and electoral processes. These adversaries also steal US defense secrets and intellectual property. They can use personally identifiable information for malign purposes, such as counter intelligence, social engineering, or ransomware attacks. China and Russia are the two greatest threats to the US. They are conducting malicious cyber campaigns to erode US military advantages, threaten US infrastructure, and reduce US economic prosperity. Cyberwarriors of the future will need to make use of offensive skills to defend US interests.

Figure 9.4 Some cyber warfare operators [17].

- *Cyber Exercises:* These have become an increasingly prominent tool in training and demonstrating capability. Military cyber exercises come in various shapes and forms. They are often run in a virtual environment. Its benefit is the organizer's ability to collect data as the activities of different teams can be more easily recorded. The short duration of military cyber exercises can impact officials' conception of the strategic potential of this space [18].

- *Military Satellites:* This technology is applied in protecting military networks and civilian networks from malicious actors and potential cyber-attacks. It is revolutionizing cybersecurity, allowing militaries to detect and respond to online threats faster than ever before. Satellites are equipped with powerful sensors that can detect cyber-related activities, such as malicious software and data breaches. This helps militaries to identify potential malicious actors and respond to them before a cyber-attack takes place. Military satellites are also being used to allow militaries better collaborate with other countries and international organizations. By using these satellites, organizations can monitor and analyze a wider area for potential threats [19]. Figure 9.5 shows a typical military satellite system [20].

Figure 9.5 A typical military satellite system [20].

9.5 BENEFITS

Cybersecurity is maintaining the security, integrity, and authenticity of information stored on electronic systems. It represents the world's fastest-growing field of defense. Government at all levels, law enforcement agencies, and military organizations face an ever-evolving battleground of potential cybersecurity threats. Cybersecurity is important for the Department of Defense (DoD) because nearly 100 percent of day-to-day operations are completed on some type of information system.

Cyber security officers are responsible for protecting military networks and the country against cyber-attacks from enemy forces. Other benefits include [20,21]:

- *Defend Military Networks:* Cybersecurity in the military serves to defend Military networks, protect US interests against cyber-attacks, and perform offensive cyber tactics in support of combat missions. For those with military experience, working in the critically important arena of cybersecurity is an opportunity to continue to serve your country, combating the ever-growing threat of cyber crime.

- *Cyber Defense:* The military sector has been taking steps to gain a better grasp of the problems of cyber defense. The overall goal of the military's cybersecurity and technology strategy is to mitigate any potential hazards. Military forces have developed cyber capabilities to aid them in combat and to secure their networks from foes in times of peace. Defense groups, such as armed forces on land, at sea, and in the air, can only execute important duties and missions if information is exchanged securely. Government agencies at the federal, state, and local level are intensifying their efforts to recruit military talent to fight the war on cyber crime.

- *Training:* Officers typically enter the military after they have completed a four-year college degree program. Enlisted service members can transition to officer positions through a variety of pathways and earn a degree while serving. Cybersecurity officers may have an education in computer science or a related field. Like other officers, they complete a comprehensive training program covering responsibilities, military structure and etiquette, traditions, and leadership development. One of such trainings is shown in Figure 9.6 [23].

Figure 9.6 A typical military training [23].

- *Rewarding Career:* An interest in cybersecurity in the military can lead to an exciting and rewarding career. This career field includes entry level, technical and professional careers related to the design, development, support and management of hardware, software, multimedia, and systems integration services. Cybersecurity professionals stop malicious actors from causing harm, damage, or disruption. They can further career opportunities by earning certifications. Cybersecurity career requires a strong programming foundation, knowledge of information systems and networks, and expertise with the latest security trends and threats.

9.6 CHALLENGES

Although cybersecurity and the military seem to have a natural partnership, there are some challenges to pursuing a career in this field. Although the military life develops many skills which are useful during periods of war and conflict, adapting those skills to a regular civilian job as an ex-military person can be very challenging.

Troops often traverse extreme terrains. Hardware needs to be lightweight enough to carry, but tough enough to withstand adverse conditions. New wearable technology captures speech, vision, location, and data to send to headquarters and fellow tactical teams, and its reliability is crucial to ensure the safety of soldiers and missions.

Current and future contracts must provide for threat-realistic, independent security assessments by the DOD of commercial clouds, to ensure critical data is protected. By knowing the limitations of the machine's algorithm, hackers can manipulate its effectiveness. Other challenges of cybersecurity in the military include the following [24-27]:

- *Expensive Training*: Cybersecurity is not exactly an entry-level field. Although earning industry-recognized certifications is critical to developing your career in cybersecurity, cost of certification is expensive. Current methods used to educate students about cyber, including annual Navy Knowledge Online training, are perceived to be ineffective.

- *Operational Complexity*: Cyber warfare often lacks the central components of war: large-scale physical destruction and massive violence. Some critics have questioned: What good is cyber in war? What is the utility of military cyber operations in conflict situations and what obstacles exist? They perceive the operational complexity of a cyber operation and the potential unsuitability of military cyber operations in war or conflict contexts. They suggest proper timing of the use of cyber capabilities.

- *Overuse of the Military:* This presents challenges for at least two reasons. The first is the practical risk of creating a "crowding-out" effect. Cyber threats are proliferating at a dramatic rate because we are making more and more use of information systems. For that reason, cybersecurity is not just something that citizens and private companies can expect to outsource to the military. Second is the risk of militarizing a major new aspect of domestic security. A potential approach for the government is to support the private sector in providing its own security.

- *Arms Race:* The cyber arms race demands constant vigilance and innovation from cybersecurity professionals. Insider threats remain a significant concern within government, military, and law enforcement agencies. These threats can come from malicious insiders who intentionally compromise systems or leak sensitive information.

- *Critical Infrastructure:* Cyber threats to critical infrastructure are of major concern. As our society becomes more intertwined with technology, the potential vulnerability of our critical infrastructure to cyberattacks intensifies. Essential systems such as power grids, water supply networks, transportation systems, and other crucial services have become targets. Any disruption to any of these infrastructures could pose significant risks to national security and public safety, making them an enticing target for cybercriminals.

- *Supply Chain:* The vulnerability of the supply chain is another major concern. As organizations increasingly interconnect and rely on third-party vendors, the security of their supply chains has become a pressing issue. Cyber criminals often exploit these third-party relationships, targeting suppliers to gain access to sensitive information or propagate malware through shared systems.

9.7 CONCLUSION

Cybersecurity is a fast-growing industry with a labor shortage, making it one of the most sought-after careers in the country. To address the shortage of cybersecurity professionals, some colleges and universities now offer a number of degrees in cybersecurity and other IT fields that you can complete at home at your own pace,. You can enroll in such an accredited program to receive your bachelor›s degree in cyber security. The military offers extensive cybersecurity training opportunities to ensure everyone knows how to best mitigate cyber risks. The Federal Virtual Training Environment (FedVTE) provides free online cybersecurity training to veterans.

Cybersecurity is not limited to the military or businesses. Everyone has information to protect. Every organization is counting on our cybersecurity experts to detect system vulnerabilities and protect sensitive data. Cyber threats are a growing concern for the military, creating a need for cybersecurity awareness and education. Cybersecurity is critical to their mission success and daily operations. The military is reaching out to even younger students through high school talent searches in the form of cybergame to identify prodigies. More information about cybersecurity in the military can be found in the books in [28-36] and the following related periodicals:

- *Journal of Cybersecurity*
- *Cyber Security Journal*
- *Security Magazine*
- *United States Cybersecurity Magazine*

REFERENCES

[1] "Study on the use of reserve forces in military cybersecurity," April 2020, https://css.ethz.ch/content/dam/ethz/special-interest/gess/cis/center-for-securities-studies/pdfs/Cyber-Reports-2020-03-military-cybersecurity.pdf

[2] M. N. O. Sadiku, C. M. M. Kotteti, and J. O. Sadiku, "Cyber security in the military," *International Journal of Trend in Research and Development,* vol. 10, no. 3, May-June 2023, pp. 422-427.

[3] N. Allen, "The importance of cybersecurity in military," October 2021, https://saltcommunications.com/news/the-importance-of-cybersecurity-in-military/

[4] D. Vergun, "DOD works to increase cybersecurity for U.S., allies," September 2020a https://www.defense.gov/News/News-Stories/Article/Article/2351916/dod-works-to-increase-cybersecurity-for-us-allies/

[5] "Eliminating the complexity in cybersecurity with artificial intelligence," https://www.wipro.com/cybersecurity/eliminating-the-complexity-in-cybersecurity-with-artificial-intelligence/

[6] M. N. O. Sadiku, S. Alam, S. M. Musa, and C. M. Akujuobi, "A primer on cybersecurity," *International Journal of Advances in Scientific Research and Engineering*, vol. 3, no. 8, Sept. 2017, pp. 71-74.

[7] M. N. O. Sadiku, M. Tembely, and S. M. Musa, "Smart grid cybersecurity," *Journal of Multidisciplinary Engineering Science and Technology*, vol. 3, no. 9, September 2016, pp.5574-5576.

[8] "FCC Small Biz Cyber Planning Guide," https://transition.fcc.gov/cyber/cyberplanner.pdf

[9] "Top 10 most common cyber attacks," https://www.edoxi.com/studyhub-detail/top-most-common-cyber-attacks

[10] Y. Zhang, "Cybersecurity and reliability of electric power grids in an interdependent cyber-physical environment," *Doctoral Dissertation,* University of Toledo, 2015.

[11] "Global military cybersecurity market – Industry trends and forecast to 2030," Unknown source.

[12] K. Gronberg, "Why military agencies must establish cybersecurity readiness now through comply-to-connect," July 2020, https://www.c4isrnet.com/opinion/2020/07/13/why-military-agencies-must-establish-cybersecurity-readiness-now-through-comply-to-connect/

[13] J. Gill, "Pentagon's cybersecurity tests aren't realistic, tough enough: Report," January 2022, https://breakingdefense.com/2022/01/pentagons-cybersecurity-tests-arent-realistic-tough-enough-report/

[14] "The U.S. army is revamping its cybersecurity approach," April 2023, https://www.tripwire.com/state-of-security/us-army-revamping-its-cybersecurity-approach

[15] "Cybersecurity for veterans," https://niccs.cisa.gov/education-training/cybersecurity-veterans

[16] "The complete cybersecurity career guide for veterans," https://bootcamp.pe.gatech.edu/blog/the-complete-cybersecurity-career-guide-for-veterans/

[17] "The opportunities and challenges of military cyber exercises," March 2022, https://www.cfr.org/blog/opportunities-and-challenges-military-cyber-exercises

[18] T. M. Cronk, "Nakasone says US works to stay ahead of cybersecurity curve," May 2021, https://www.defense.gov/News/News-Stories/Article/Article/2638552/nakasone-says-us-works-to-stay-ahead-of-cybersecurity-curve/#:~:text=%22We're%20all%20here%20with, meet%20the%20critical%20challenges%20ahead.%22

[19] M. Frąckiewicz, "Military satellites and the future of cybersecurity," https://ts2.space/en/military-satellites-and-the-future-of-cybersecurity/

[20] K. Drummond, "Darpa: Use NASCAR parts to rev up satellites," March 2012, https://www.wired.com/2012/03/darpa-satellites/

[21] "Cybersecurity and information technology," https://www.todaysmilitary.com/careers-benefits/career-fields/cybersecurity-and-information-technology#:~:text=They%20serve%20to%20defend%20Military,including%20the%20Reserve%20and%20Guard.

[22] M. Moore, "Why cybersecurity is a great career option for military veterans," https://onlinedegrees.sandiego.edu/why-cyber-security-is-a-great-career-option-for-military-veterans/

[23] G. Harkins, "Lawmakers are worried COVID-19 did long-term damage to military training," April 2021, https://www.military.com/daily-news/2021/04/12/lawmakers-are-worried-covid-19-did-long-term-damage-military-training.html

[24] "How to make the jump from the military to the cybersecurity field," April 2022, https://www.mastersindatascience.org/resources/military-cybersecurity-field/

[25] M. Schulze, "Cyber in war: Assessing the strategic, tactical, and operational utility of military cyber operations," https://ccdcoe.org/uploads/2020/05/CyCon 2020 10 Schulze.pdf

[26] I. Wallace, "The military role in national cybersecurity governance," https://www.brookings.edu/articles/the-military-role-in-national-cybersecurity-governance/

[27] "The challenges for government, military, and law enforcement cybersecurity," May 2023, https://truefort.com/law-enforcement-cybersecurity/#:~:text=This%20cyber%20arms%20race%20demands%20constant%20vigilance%20and%20innovation%20from%20cybersecurity%20professionals.&text=Insider%20threats%20remain%20a%20significant,systems%20or%20leak%20sensitive%20information.

[28] F . Liu et al., *Science of Cyber Security.* Springer, 2018.

[29] L. Currie-McGhee, *Intelligence and Cybersecurity in the Military (Careers in the Military).* Referencepoint Pr Inc., 2022.

[30] D. V. Puyvelde and A. Brantly (eds.), *US National Cybersecurity: International Politics, Concepts and Organization (Routledge Studies in Conflict, Security and Technology).* Routledge, 2017.

[31] A. B. Lowther and P. A. Yannakogeorgos, *Conflict and Cooperation in Cyberspace: The Challenge to National Security.* Boca Raton, FL: CRC Press, 2016.

[32] U. S. Military, *Department of Defense (Dod), and A. Bardwell, Cybersecurity Education for Military Officers - Recommendations for Structuring Coursework to Eliminate Lab Portion and Center Military-Relevant Discussions on Cyber-Defense Management.* Amazon Digital Services LLC, 2018.

[33] J. L. Caton and Strategic Studies Institute, *Examining the Roles of Army Reserve Component Forces in Military Cyberspace Operations.* Independently Published, 2019.

[34] D. Snyder, *Improving the Cybersecurity of U.S. Air Force Military Systems Throughout Their Life Cycles.* RAND, 2015.

[35] P. Shakarian, J. Shakarian, and A. Ruef, *Introduction to Cyber-Warfare: A Multidisciplinary Approach.* Elsevier Science, 2013.

[36] J. Arquilla, *Bitskrieg: The New Challenge of Cyberwarfare.* Polity Press, 2021.

CHAPTER 10

BIOTECHNOLOGY IN THE MILITARY

"A nation's ability to fight a modern war is as good as its technological ability."
– Frank Whittle

10.1 INTRODUCTION

Biotechnology (or biotech) is a combination of biology and technology. So biotech is basically technology based on biology. Biotech involves not just a single technology but a wide range of technologies that share two key characteristics: working with living cells and having a wide range of uses that can improve our lives. It embraces a wide range of procedures for modifying living organisms to suit human purposes. Related fields include molecular biology, cell biology, microbiology, biochemistry, bioengineering, biomedical engineering, biomanufacturing, molecular engineering, and bioinformatics. The multidisciplinary nature of biotechnology is illustrated in Figure 10.1 [1]. The demand for new drugs is a major incentive for the explosive growth in the biotechnology industry.

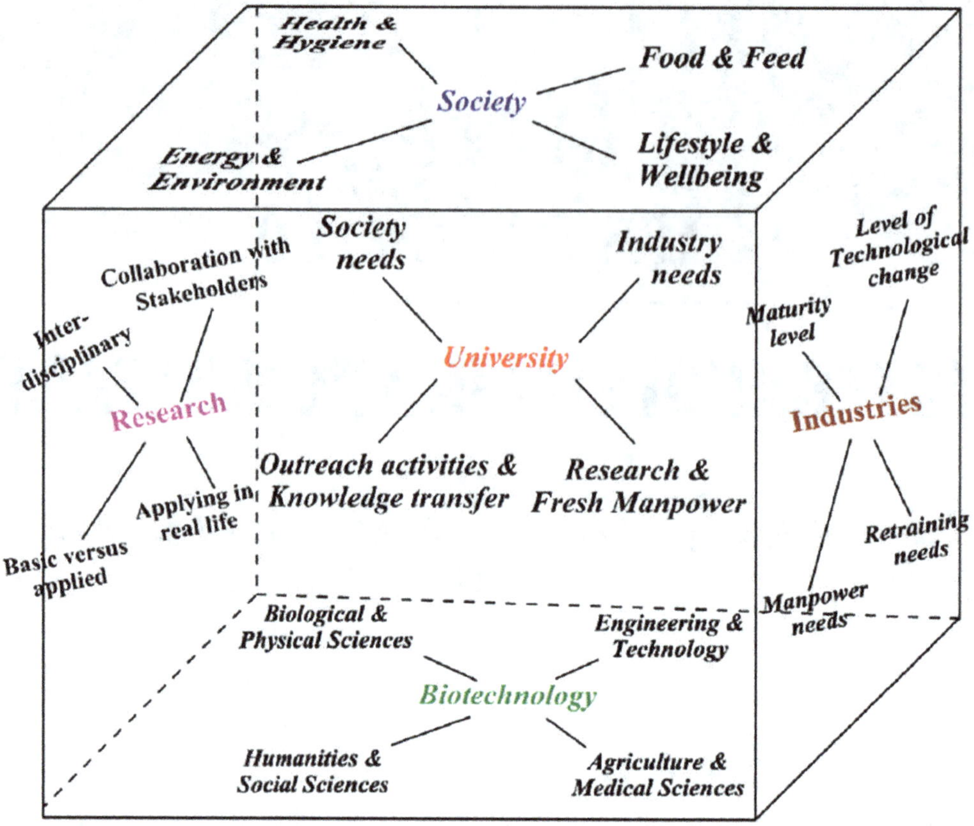

Figure 10.1 The multidisciplinary nature of biotechnology [1].

Although the term "biotechnology" was not used until 1919, ancient civilizations used biological processes to leaven bread, brew beer, and ferment wine. There is no substance as important as deoxyribonucleic acid (DNA), because it carries the hereditary information that determines the structures of proteins. Once enzymes could be isolated, scientists could begin to direct the recombination of DNA and perform genetic engineering, which is the basis of the biotechnology industry. Modern biotechnology began in the 1970s after the development of genetic engineering that enabled scientists to modify the genetic material of living cells. Since the 1970s, when the first commercial company was founded to develop genetically engineered products, the biotechnology industry has grown rapidly. Knowledge of DNA sequences has become indispensable for biotech. All organisms are made up of cells that are programmed by the same basic genetic material, called DNA (deoxyribonucleic acid). DNA is made up of four similar nucleotides (designated by an organic base they contain, abbreviated A, T, C, and G). A genome is all the DNA in an organism, including all of its genes [2]. Biological sensing and analysis capabilities may extend to monitoring the health, safety, and performance of soldiers in the field. Biotechnology will have a large impact on the defense sector.

Our world is undergoing a military revolution characterized by electronics, computers, communications, and microinformation technology. Biotechnology is developing quite rapidly and has had an enormous effect on the progress of science and technology. It is a broad term used to describe technological innovation based on biology. It covers all aspects of living organisms, from medicine to agriculture. It is rapidly changing and growing. Biotechnology and the military are teaming up to help soldiers meet the challenges of needing quick mobility and different environmental background. Biotechnology offers novel opportunities for improving warfighter survivability on the battlefield [3].

This chapter examines the various uses of biotechnology in the military. It begins with explaining what biotechnology is all about. It discusses military biotechnology and provides some of its applications. It covers military biotechnology around the world. It highlights the benefits and challenges of military biotechnology. The last section concludes with comments.

10.2 WHAT IS BIOTECHNOLOGY?

Biotechnology is technology that utilizes biological systems or living organisms to develop products. It is basically applied biology that forms the interface between biology and engineering. It is the one of the most rapidly growing field of technology today.

Biotechnology pervades almost all aspects of our daily life; it affects the foods we eat, the safety of the water we drink, the clothes we wear, the medications we take, etc.

Biotechnology can roughly be divided into three main parts [4]:

- *Green Biotechnology*: This involves agricultural processes. The foundation of green biotech is crop improvement and production of novel products in plants.
- *Red Biotechnology*: This involves healthcare processes. It uses the human body's own tools and weapons to fight diseases.
- *White Biotechnology*: This field is connected with industry and environmental processes. Most of the white biotech processes results in the saving of water, energy, chemicals and in the reduction of waste.

Popular biotech fields include [5]:

- *Genetic engineering*: This is the direct manipulation of DNA molecules to produce modified plants, animals, or other organisms using biotechnology. Through genetic engineering, organisms can be given targeted combinations of new genes.
- *Tissue culture*: This is a method where by fragments of tissue from an animal or plant are transferred to an artificial environment to continue to survive and function.
- *Cloning*: This describes the process (of breeding) used to create an exact genetic replica of another cell. There are three different types of cloning: (a) Gene cloning, which creates copies of genes, (b) Reproductive cloning, which creates copies of whole animals, (c) Therapeutic cloning which creates embryonic stem cells.

Humankind has used biotechnology in several areas such as agriculture, healthcare, environment, and industrial biotech. The relationships between these various applications are shown in the biotechnology tree in Figure 10.2 [6]. Figure 10.3 shows the classification of biotechnology sector [7].

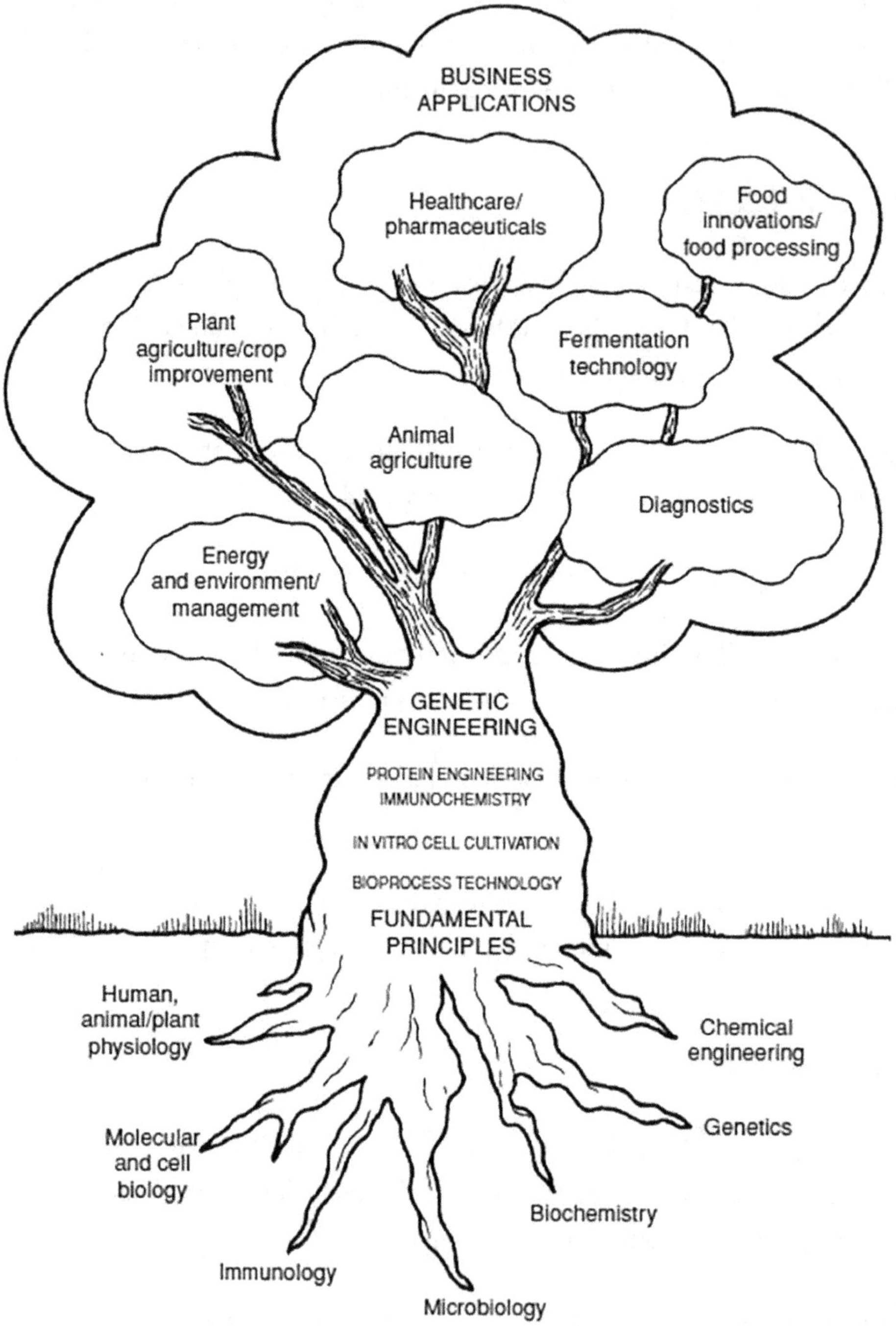

Figure 10.2 The biotechnology tree [6].

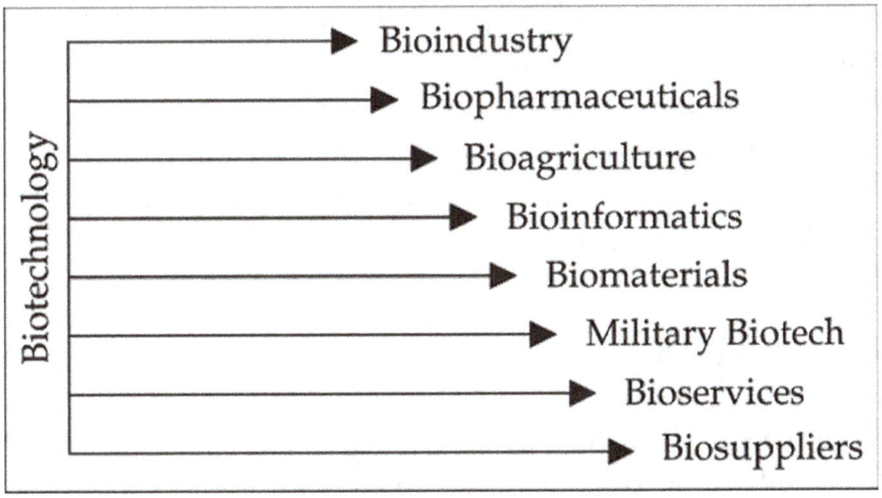

Figure 10.3 The classification of biotechnology sector [7].

10.3 MILITARY BIOTECHNOLOGY

Biotechnology is an interdisciplinary field that combines biology, genetics, and engineering to develop solutions for defense challenges. For military affairs, biotechnology is developing at a rapid pace. It is playing a critical role in medical protection. Biotechnology and the military are strengthening the power of soldiers and resisting fatigue. It senses and monitors the battlefield. Human bodies could experience direct-effect weapons that would alter their biological features because biotechnology looks at a molecular level. Due to this, scientists can soon unlock ways to control, reconstruct, design, and simulate molecules in humans. Scientists could change cell functions as needed with the interaction of proteins. Figure 10.4 shows the 2024 convention on biotechnology [8].

Figure 10.4 The 2024 convention on biotechnology [8].

The knowledge derived from the study of the genetic characteristics, molecular biology, metabolism, and biology of organisms promises to facilitate the design of devices, software, and genetically altered organisms. Modern biotech research has a focus on life structure's microcosmos. It is playing a critical role in the diagnosis and prevention of diseases, protection from biochemical toxic agents, and treatment of war injuries. Weapons created by biotechnology would be more destructive than conventional methods of the past, such as nuclear weapons and gunpowder.

10.4 APPLICATIONS OF MILITARY BIOTECHNOLOGY

Biotechnology is an engineering discipline that uses living systems to create a wide range of products. The technology can be used to produce an enormous range of things from food and medicines to textiles and fuels. It has many applications in the military. Although applications of military biotechnology are complicated, the finished products are convenient to carry, easy to use, and do not require large support systems. These applications that can support the rapidly emerging bioeconomy which can have significant benefits for national security. Specific applications include the following [9-11]:

1. *Defense and Attack:* Future combat systems of all types will be affected by biotechnology. Biotechnology can be used to develop new weapons systems that can be used for defense and attack. It is helping to reach all new levels of combat and helping large armies defend nations against adversaries of all sizes. The warfighter will undoubtedly be impacted directly by biotechnology innovations, including capabilities specifically related to improving force health protection and mission readiness. Optimizing warfighter performance will also include enhanced abilities to sense the environment. Combat functions themselves may be modeled on natural, efficient biological processes.

2. *Treatment of War Injuries:* Biotechnology can be used to treat war injuries. Modern biotechnology has played an important role in treatment of war injuries, prevention and diagnosis of diseases, and protection. It is playing a critical role in the diagnosis and prevention of diseases, protection from biochemical toxic agents, and treatment of war injuries. Another game-changer could come from how soldiers heal in battle after experiencing an injury. Artificial skin could insulate soldiers from environmental extremes, as well as provide frontline treatment for wounds. Figure 10.5 shows a wounded soldier [21].

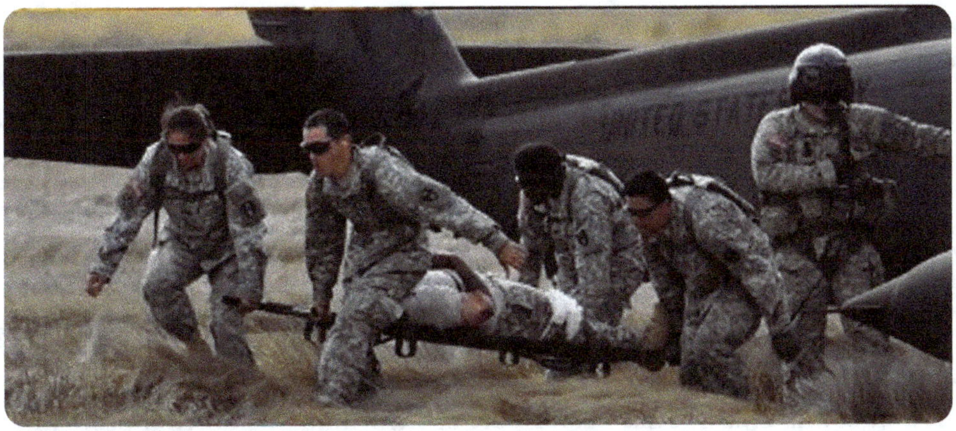

Figure 10.5 A wounded soldier [12].

3. *Materials and Equipment:* Biotechnology uses biological processes, cells or cellular compounds to develop new materials, products, and technologies. It can be used to develop new materials and equipment for military platforms and infrastructure. It leverages unique properties of materials for military platforms and infrastructure, including those that are stronger, lighter, self-healing, less toxic, more efficient, and/or faster to manufacture than current alternatives. These materials can be stronger, lighter, self-healing, less toxic, more efficient, and/or faster to manufacture. Biomaterials could be used to develop new reagents for next-generation explosives, harvest rare earth materials, enhance armor protection, and develop specialized bio resins and polymers that offer increased performance in various applications. Modern biotechnology continues to advance the manufacture of bioderived materials through techniques such as enzyme engineering, cell-free reactions, and the expansion of classic fermentation-based technologies. Biological systems offer endless possibilities for the military to model synthetic materials. Figure 10.6 shows a scientist holding biosensing materials in a laboratory [10].

Figure 10.6 A scientist holding biosensing materials [10].

4. *Monitoring and Detection*: Biotechnology can be used to detect, identify, and monitor chemical, biological, radiological, and nuclear threats. Biotechnology can be used to sense and monitor the battlefield. Soldiers in the future will wear or carry sensors that can detect signature molecules in the environment, alerting them to changes that may be caused by enemy activity or influence. Sensors may provide early warning of an enemy intention to pollute the battlefield with chemical or biological agents.

5. *Biomanufacturing:* Biomanufacturing is the use of biological mechanisms in the manufacturing process. It is a process in which organisms and their biological systems are used to produce chemicals and biomaterials. It has been a part of the military industrial base since World War I. In its infancy, biomanufacturing relied on the availability of naturally occurring microbial strains that produced specific materials. Modern biomanufacturing combines a variety of disciplines, including engineering, biology, chemistry, and computer science, and facilitates the production of biologically derived materials on a commercial scale. Advancements in the fields of synthetic biology, artificial intelligence, and robotics have resulted in the rapid expansion of small-scale production capabilities. The processing of biologically derived materials is poised to revolutionize the way civilian and military sectors produce materials. The operations of a biomanufacturing plant can be separated into two phases: fermentation and product recovery. In the fermentation phase, sugars and other nutrients are converted into biomass and the desired product. In the product recovery phase, the alcohol is distilled and purified. Companies that focus on using fermentation to make products optimize both process development and strain performance. Building and operating large biomanufacturing facilities involve notable financial risk. Work is rapidly advancing that would introduce biomanufacturing processes for production of fuels, chemicals, and even construction materials. DoD develops biomanufacturing at home and with allies and partners to create a self-sustaining domestic biomanufacturing ecosystem.

6. *Supply Chains:* The supply chains have become burdensome and unwieldy, plaguing defense acquisitions. Biotechnology can make a significant contribution to addressing vulnerabilities in Department of Defense (DOD) supply chains. Recognizing its potential to revitalize supply chains, DOD recently named biotechnology as one of its modernization priorities. Defense planners must envision biotechnology products from inception through their full development and manufacture pathways that allow these technologies to be successfully shepherded. Initiatives from both the White House and DOD now promote the establishment of domestic supply chains that use biotechnology- based materials and biochemicals for high-value chemical precursors, military armor, energetics, and propellants.

10.5 MILITARY BIOTECHNOLOGY AROUND THE WORLD

Warfare has historically been the driving force for technology with numerous nations around the globe striving to develop the most advanced and innovative armaments to protect their domestic and international interests. A greater degree of visibility into the various biotechnology applications is being developed across governments and military organizations worldwide. Global military spending continues to rise year after year and the trend is not expected to stop any time soon. We consider how some nations are using biotechnology in their defense sector.

- *United States:* The US military maintained superiority in the area of science and technology for many decades. America's competitive edge makes it challenging for the US to deliver critical technologies. The Defense Advanced Research Projects Agency (DARPA) has investigated the behavior of insects and other animals in research for the Department of Defense (DOD). The principles of design, biosynthesis, and structure-property correlations in "living" materials and systems will be very important in determining new military applications of biotechnology. Thinking in terms of biological systems may not only provide solutions to specific problems, but may also provide clues to future opportunities. The US Army has declared biomimetics one of its primary focus areas for basic research [13]. Today there is a growing number of organizations that directly consider biotechnology capabilities and the issues that can affect warfighters. can affect the warfighter indirectly. The DoD must maintain a strong biodefense program to address the risks of deliberate use of biological weapons, with focus on great power competition with advanced adversaries China and Russia. Figure 10.7 depicts the collaboration between the Army and academic institutions [14].

Figure 10.7 Collaboration between the Army and academic institutions [14].

- *Canada:* A Canadian company is working on "Quantum Stealth." They say they can hide where troops, artillery, tanks, and even buildings are hiding. It is not just the military that uses stealth. Predators in the animal kingdom rely on stealth to attack their prey. Asymmetric warfare is more suitable for cloaking or stealth technology. You could make another argument that stealth technology could lead to more peace. This is because smaller nations would not want to antagonize a power with a large military.

- *China:* China has spurred a significant increase in biotechnology research and development, with an anticipated increase of seven percent per year between 2021 and 2025. More specifically, China has made efforts to acquire international data that can facilitate assessment and control of health care for different countries. China›s strategic investments in the United States are relevant as well. International companies that build facilities in the United States from the ground up are not subject to scrutiny. Although the construction, associated tax base, and potential job creation can be appealing locally, the risk to national security could well go unnoticed and unregulated. Consequently, near peer competitors could gather data about US technologies and citizens without being noticed. China has conducted «human testing» on members of the People›s Liberation Army in hope of developing soldiers with «biologically enhanced capabilities,» thereby creating biologically enhanced super soldiers [15]. PLA strategists believe that achieving "mental dominance" will be critical in future military competition across the spectrum from peacetime to warfighting. But Western scientists consider it unethical to seek to manipulate genes to boost the performance of healthy people. The People›s Republic of China poses the greatest threat to America today. There are no ethical boundaries to Beijing›s pursuit of power. Figure 10.8 shows some Chinese soldiers [16], while Figure 10.9 shows the asymmetry in ethics that exists between the West and China [17].

Figure 10.8 Some Chinese soldiers [16].

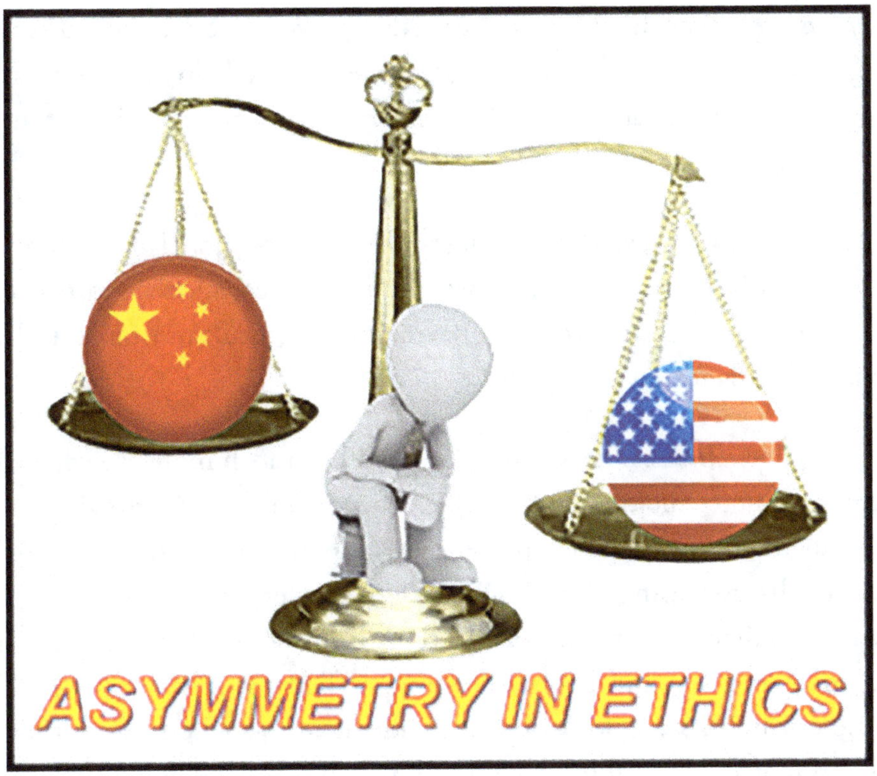

Figure 10.9 Asymmetry in ethics that exists between the West and China [17].

- *Russia:* Russia continues to invest heavily into human enhancement technologies (BHE). It has also led efforts to undermine global norms against the proliferation and use of weapons of mass destruction. Russia has dangerously increased the spread of disinformation about biological and chemical weapons, including during the war against Ukraine. There are concerns that Russia is considering further use of chemical or biological weapons in the future [18].

10.6 BENEFITS

Humaneness in the conduct of war has become the focus of attention recently, and weapons of mass destruction are banned to reduce casualties. Modern biotechnology makes it possible to combine two or more pathogenic genes and place them inside a susceptible living body to create a multiple-vulnerating effect. Future biobased technologies will undoubtedly create applications that cannot be predicted now. Other benefits of biotechnology in the military include the following:

- *Precision:* Biotechnology allows us attack at a more precise level. Injuries might be limited to a specific gene sequence or a specific protein structure. Through gene manipulation, we can attack or injure one or more key human physiological functions (the ability to learn, memorize, keep one's balance, or even act aggressively) without a threat to life. Precision injury and damage are two vulnerating methods based on genomics and proteomics. They are completely different from traditional weapons of war that directly damage tissues and organs. Thus, current military biotechnology possesses a quality of "mercy."

- *Prevention of Diseases*: Biotechnology can be used to prevent and diagnose diseases. It can be used to protect against biochemical toxic agents. It can be used to help soldiers resist fatigue. Modern biotechnology maintains a rapid pace of development and plays an important role in medical protection.

- *Improving Materials:* Biological systems might also serve as models for improving materials for uniforms, particularly by reducing their weight and increasing their functionality. A soldier's clothing must protect against extremes of weather, chemical and biological agents, heat and humidity, and other factors.

- *Feeding Growing Populations:* The world in 2025 will be much more crowded, and resources will be at a premium. Biotechnology will provide a means of feeding growing populations. Many foods will be engineered to provide optimal nutrition and minimize spoilage. Therapeutics for treating chronic diseases using biotechnology-derived methods and products will be common. Vaccines will be available against most infectious diseases. With a better understanding of the basis of life, many of the painful conditions that afflict mankind in 2000 will be preventable. Although soldiers in 2025 will look outwardly identical to soldiers today, they will be stronger, have longer endurance, and will be more resistant to disease and aging.

- *Medical Treatments*: Biotechnology offers advanced medical treatments for battlefield injuries. It enables personalized medicine for healing wounds and treating specific injuries or diseases based on a soldier's genetic profile. It also enables rapid development of vaccines against emerging pathogens and antidotes for chemical and biological warfare agents.

10.7 CHALLENGES

Every organization that works with biotechnology faces unique challenges and needs. The challenges include the following:

- *Dual Use:* Biotechnology is inherently dual use, meaning it could be used both for legitimate and nefarious purposes. While the opportunities for combatting disease, cleaning up environmental pollutants, and harnessing scarce natural resources are positive outcomes, we should also be aware of the challenges and risks such as a rogue actor developing offensive biological warfare capabilities or the weaponization of pathogens to harm fragile biological ecosystems. As biotechnology advances and proliferates, the DoD will need to keep track of how various capabilities could be used for nefarious purposes, including deliberate attacks against populations and deployed forces.

- *Risk:* Biotechnologies may also be used in ways that pose risks to our armed forces, societies, and the environment. There are proliferation risks of new types of bioweapons created from accessible biotechnology research, including as fueled by generative AI. There is also unpredictable spread of biological agents with potentially irreversible impacts.

- *Fear:* As emerging technologies become more accessible, many in the Pentagon are concerned that adversaries might soon challenge or surpass American strengths. The democratization of biotechnology can enable adversaries to achieve technological parity. Some global competitors may be poised to achieve superiority.

- *Lack of Standardization:* Process development schemes are often considered lucrative trade secrets, but a lack of standardization in industrial process requirements continues to mean that ad hoc development can be an approach fraught with risk. Developing and deploying new biotechnologies for military use often faces stringent regulatory hurdles at both national and international levels.

- *Ethical Concerns:* There could be potential misuse of biotechnology for offensive purposes, such as genetically engineered weapons. This raises ethical concerns that should be addressed. Open communication and transparency are crucial to manage public concerns regarding military applications of biotechnology.

- *Biosecurity*: Strict regulations and safeguards must be implemented to prevent the proliferation of dangerous biological agents. Strict protocols and safeguards are necessary to prevent accidental release of dangerous pathogens during research, development, and deployment.

- *Unintended Consequences:* Modifying biological organisms for military purposes could lead to unforeseen ecological impacts if released into the environment, potentially causing unintended harm to ecosystems and human populations.

10.8 CONCLUSION

Biotechnology involves the manipulation of living organisms or their components to produce useful products. It includes the manufacture of products ranging from food-grade sweeteners to fuel alcohol, as well as the use of chemicals to modify the behavior of biological systems. Biotechnology is still an emerging field. Biotechnology and the military have exciting possibilities. Although biotechnology is one of the most versatile, exciting, and innovative technologies of the 21st century, its benefits for defense have yet to be fully explored. DoD seeds opportunities in biotechnology as part of its broader responsibility to ensure our enduring technological advantage, with significant implications for national security and economic competitiveness.

Biotechnology may be reaching a critical junction. As it continues to mature, proactive policy becomes necessary for the federal government to leverage emerging capabilities effectively and remain competitive. For more information about biotechnology in the military, one should consult the books in [6,13,19] and the following related journals devoted to robotics:

- *Military Review*
- *Journal of Military Learning*
- *NCO Journal*

REFERENCES

[1] B. P. R. Narasimharao, "Biotechnology education and societal demands: Challenges faced by biotechnology and human resources development," *Social Responsibility Journal*, vol.6, no. 1, 2010, pp 72-90.

[2] M. N. O. Sadiku, T. J. Ashaolu, S. M. Musa, "Biotechnology: A primer," *International Journal of Engineering Research and Advanced Technology,* vol. 4, no. 9, Sept. 2018, pp. 87-91.

[3] M. N. O. Sadiku, P. A. Adekunte, and J. O. Sadiku, "Biotechnology in the Military," *International Journal of Trend in Scientific Research and Development*, vol. 8, no. 5, September-October 2024, pp. 226-235.

[4] "Managing innovations in biotechnology: European project semester 2006," UK Presidency of the European Union 2005; the Lisbon Agenda. (Unknown Source)

[5] Z. Naz, "Introduction to biotechnology," November 2015 https://www.researchgate.net/publication/284169166

[6] J. E. Smith, *Biotechnology*. Cambridge, UK: Cambridge University Press, 5th edition, 2009, p. 16.

[7] https://www.researchgate.net/figure/Classification-of-Biotechnology-Sector fig1 272243632

[8] T. T. Parish, "USAMMDA team joins world's biotechnology industry leaders for annual convention in San Diego," https://usammda.health.mil/index.cfm/public affairs/articles/2024/USAMMDA team joins worlds biotechnology industry leaders for annual convention in san diego#:~:text=Team%20members%20with%20the%20U.S.,June%203%2D6%2C%202024.

[9] H. S. Gibbons and A. M. Crumbley, "Accelerating transition of biotechnology products for military supply chains," July 2024, https://ndupress.ndu.edu/Media/News/News-Article-View/Article/3837475/accelerating-transition-of-biotechnology-products-for-military-supply-chains/

[10] D. DiEuliis, "Biotechnology for the battlefield: In need of a strategy," November 2018, https://warontherocks.com/2018/11/biotechnology-for-the-battlefield-in-need-of-a-strategy/

[11] "Biotechnology and today's warfighter," October 2022, https://www.rand.org/pubs/commentary/2022/10/biotechnology-and-todays-warfighter.html

[12] M. Luciani, "DARPA opens a biotechnology division," April 2014, https://english.netmassimo.com/2014/04/03/darpa-opens-a-biotechnology-division/

[13] National Research Council, *Opportunities in Biotechnology for Future Army Applications.* National Academies Press, 2001.

[14] J. P. Brayboy, "Army recognizes institute's biotechnology advances with $48 million renewal," February 2014, https://www.army.mil/article/120391/army recognizes institutes biotechnology advances with 48 million renewal

[15] K. Dilanian, "China has done human testing to create biologically enhanced super soldiers, says top U.S. official," December 2020, https://www.nbcnews.com/politics/national-security/china-has-done-human-testing-create-biologically-enhanced-super-soldiers-n1249914

[16] "China conducts biological test on its army to create super-soldiers, says top US official," December 2020, https://www.businesstoday.in/latest/world/story/china-conducts-biological-test-on-its-army-to-create-super-soldiers-says-top-us-official-280973-2020-12-09

[17] "Summary of NATO's biotechnology and human enhancement technologies strategy," April 2024, https://www.nato.int/cps/en/natohq/official texts 224669.htm#:~:text=Biotechnology%20and%20human%20enhancement%20technologies%20(BHE)%20will%20transform%20our%20economies,develop%20new%20products%20and%20technologies.

[18] "456. China: Leader in military application of biological human performance enhancement by 2030," August 2023, https://madsciblog.tradoc.army.mil/456-china-leader-in-military-application-of-biological-human-performance-enhancement-by-2030/

[19] R. L. Paarlberg, *Starved for Science: How Biotechnology Is Being Kept Out of Africa.* Harvard University Press, 2008.

CHAPTER 11

NANOTECHNOLOGY IN THE MILITARY

"Nanotechnology is manufacturing with atoms."
– William Powell

11.1 INTRODUCTION

Technology has always been key to winning wars. Wars are often won by the force with the greatest technological advantage. For this reason, technology is a central focus of defense industry research. The military has always been at the forefront of technology. Technological advances have always been an effective way to give one army a winning edge over the other. Military innovation, even in peacetime, has become an extension of war itself. Military research, however, is more than just going to war; it also enables technological advances outside of the military. For example, military research has created drones, stealth aircraft, and an improved understanding of composite materials.

Nanotechnology signifies the research and development (R&D) of technology at the nanoscale. The application of nanotechnology, specifically in the area of warfare and defense, has paved the way for future research in the context of weaponization. Advancements in this area have led to the development of nano-weapons, small robotic machines, hyper-reactive explosives, and electromagnetic super-materials. Nanotechnology represents a significant opportunity for the military to enhance the way soldiers are equipped to fight. Some potential military applications of nanotechnology are already quite advanced, and will come into play much sooner than others. For example, many sensors have already been developed which take advantage of the unique properties of nanomaterials to become smaller and more sensitive, compared to conventional technology. Portable, efficient sensors will be highly valuable to military field operatives.

Nanotechnology is the study and manipulation of matter at incredibly small sizes. It is used across all scientific and engineering fields. Nanotechnology research and development has a great role to play in materials and systems for military use. It has been a boon with its many military applications, and research conducted for military purposes can have great benefits in civilian contexts as well. The potential military applications of this technology are vast, from creating lightweight and durable armor to developing medical solutions to rapidly heal injuries on the battlefield [1].

This chapter focuses on the use of nanotechnology in various military applications. It begins with explaining what nanotechnology is all about. It discusses military nanotechnology and provides some of its applications. It covers military nanotechnology around the world. It highlights the benefits and challenges of military nanotechnology. The last section concludes with comments.

11.2 WHAT IS NANOTECHNOLOGY?

Technologies impact every aspect of our modern society. There are many ways in which our society and technology are interlinked. Nanotechnology has the potential to provide huge benefits, just like any useful technology.

The term "nano" means something small, tiny, and atomic in nature. The application of the term in science led to a field called nanotechnology. Nanotechnology refers to the characterization, fabrication and manipulation of structures, devices or materials that have one or more dimensions that are smaller than 100 nanometers. It may be regarded as an area of science and engineering where phenomena that take place at the nano-scale (10-9m) are utilized in the design, production, and application of materials and systems. It is an emerging area of that integrates chemistry, biology, and materials science to create new properties that can be exploited to gain new market opportunities [3].

Nanotechnology deals with the characterization, fabrication, and manipulation of biological and nonbiological structures smaller than 100 nm. Dimensions between approximately 1 and 100 nanometers are known as the nanoscale. As indicated in Figure 11.1 [3], the nanoscale is so small that we cannot see it with a light microscope. It is the scale of atoms and molecules. Nanotechnology involves the creation and application of materials and devices at the level of molecules and atoms. It may be regarded as the science that is conducted, researched, investigated, and experimented at the nanoscale. Nanotechnology is a multi-disciplinary field that includes biology, chemistry, physics, material science, and engineering. It is the science of small things—at the atomic level or nanoscale level. The past three decades has witnessed an increased interest and funding in nanotechnology. This has led to rapid developments in all areas of science and engineering [4].

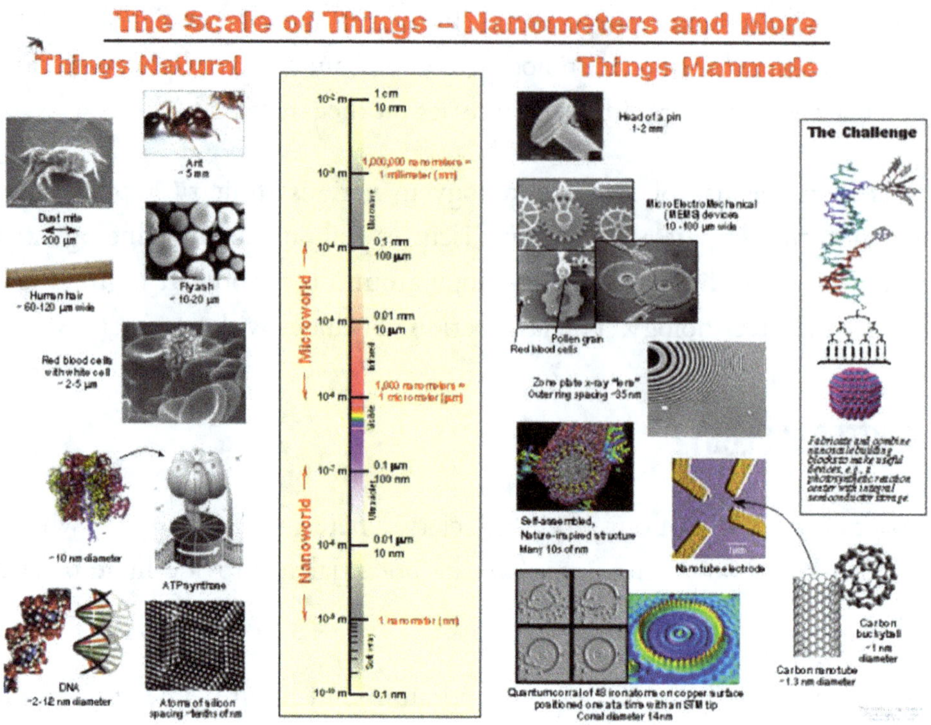

Figure 11.1 Indicating the relative scale of nanosized objects [3].

Richard Feymann, the Nobel Prize-winning physicist, introduced the world to nanotechnology in 1959 and is regarded as the father of nanotechnology. Nanotechnology involves the manipulation of atoms and molecules at the nanoscale so that materials have new unique properties. Nanomaterials are expected to have at least one dimension (length, width, height) at the nanoscale of 1 – 100 nm. One nanometer is a billionth of a meter, too small to be seen with a conventional lab microscope. Nanomaterials include nanofilters, nanosensors, nano photocatalysts, and nanoparticles. Nanomaterials are known as nanoparticles when they have nanoscale length, width, and height. Figure 11.2 portrays a technique for the preparation of nanoparticles [5].

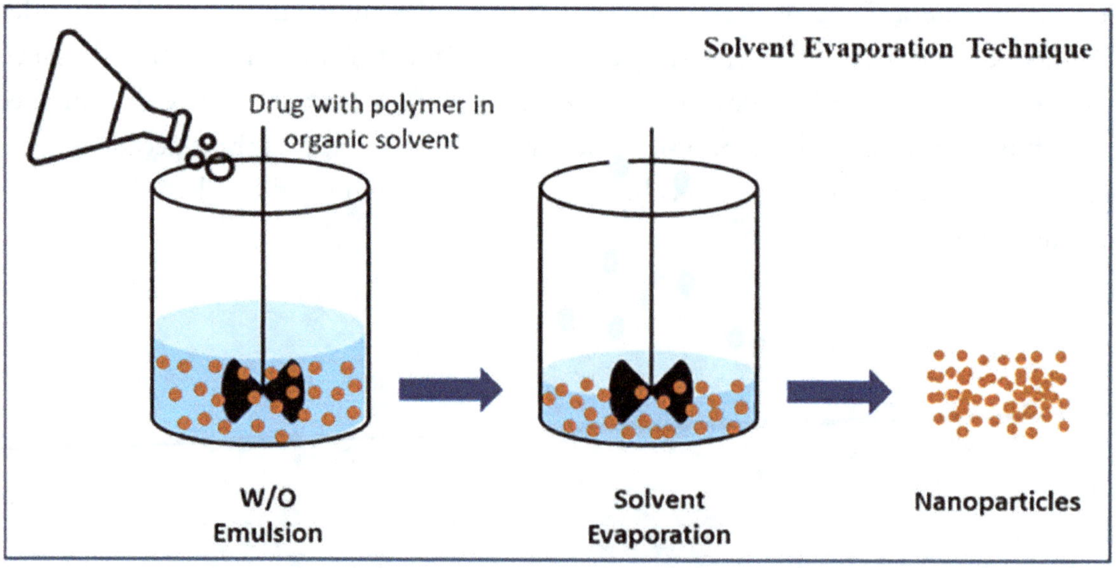

Figure 11.2 A technique for the preparation of nanoparticles [5].

Today, nanotechnology is part of our daily lives. Nanotechnology will leave virtually no aspect of life untouched. Its usages include everything from safer food processing to more efficient drug-delivery systems to powerful computer chips. Three steps to achieving nanotechnology-produced goods are [6]:

1. Scientists must be able to manipulate individual atoms.
2. Next step is to develop nanoscopic machines, called assemblers, that can be programmed to manipulate atoms and molecules at will.
3. In order to create enough assemblers to build consumer goods, some nanomachines called replicators, will be programmed to build more assemblers.

Nanotechnology is trending among scientists and engineers. Here are some underlying trends one should look for [7]:

1. *Stronger Materials*: The next generation of graphene and carbon devices will lead to even lighter but stronger structures.
2. *Scalability of Production*: One big challenge is how to produce nanomaterials that make them affordable. Limited scalability often hinders application.
3. *More Commercialization*: In addition to transforming the automotive, aerospace, and sporting goods fields, nanotechnology is facilitating so many diverse improvements: thinner, affordable, and more durable.
4. *Sustainability*: One main goal of the National Nanotechnology Initiative, a US government program coordinating communication and collaboration for nanotechnology activities, is to find nanotechnology solutions to sustainability.
5. *Nanomedicine*: There will be a mindboggling impact of nanotechnology on medicine, where advances are being made in both diagnostics and treatment areas.

Applications of nanotechnology are found in a wide range of industries, including engineering, medicine, microelectronics, manufacturing, biology, chemistry, energy, and agriculture, and life sciences. Figure 11.3 shows some applications of nanotechnology [8]. Although nanotechnology has been successfully applied in various industries, its use in the oil and gas sector is still limited.

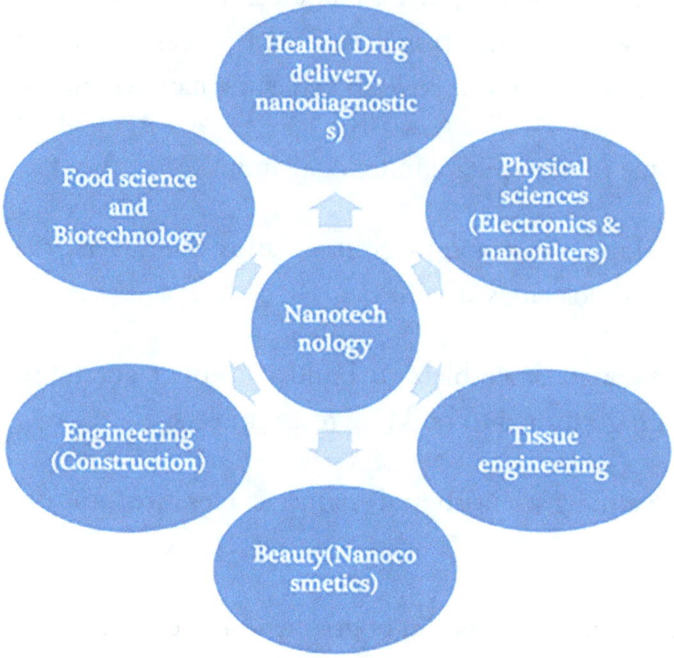

Figure 11.3 Some applications of nanotechnology [8].

11.3 MILITARY NANOTECHNOLOGY

Nanotechnology is the conception, formation, design and application of valuable materials, instruments, and systems through the maneuvering of matter on minute scale. The use of nanotechnology in the area of warfare and defense has been rapid and expansive. Military applications of nanotechnology range from improved materials for soldiers' personal protection to sophisticated sensors capable of detecting chemical and biological threats. The US Department of Defense (DoD) nanotechnology program is grouped into seven program component areas, shown in Figure 11.4 [9].

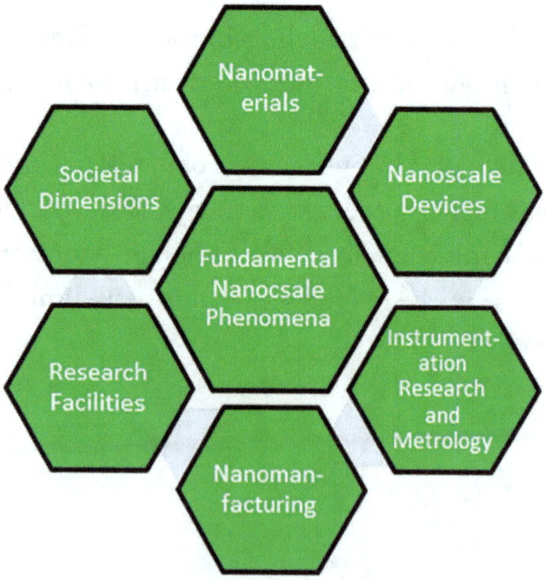

Figure 11.4 DoD nanotechnology program component areas [9].

Robotic drones are already operated by remote control. Advances in nanotechnology will allow both robotic systems and control systems to become smaller and more effective. It appears likely that most military technology will be dependent on nanomaterials. Some of the more potential applications in this area include [9,10]:

- Nano-machines to mimic human muscle action in an exoskeleton
- Stealth coatings
- Self-healing (self-repair) material
- Smart skin materials
- Adaptive camouflage
- Adaptive structures
- Increased surveillance for better protection
- Smaller cameras
- Cheap, small, and more effective weapons
- Exploration of the oceans
- Augmenting human performance
- Scratch resistant surfaces
- Stronger, thinner, and cheaper glass
- Coatings that do not degrade
- Lighter, faster aircraft which use less fuel
- Submarines and planes that can go undetected by radar
- Faster intensive medical help
- Nano-size umbrellas

11.4 APPLICATIONS OF MILITARY NANOTECHNOLOGY

Nanotechnology has numerous military applications. The most obvious are in materials science. Some other applications of nanotechnology include medicine, biological and chemical sensors, explosives, electronics for computing, power generation and storage, and structural materials for making vehicles, coatings, filters, and fabrics, and many of these applications have non-military uses as well. Some of these applications are shown in Figure 11.5 [11]. The main goals of military research into nanotechnology are to improve medical and casualty care for soldiers, and to produce lightweight, strong and multi-functional materials for use in clothing.

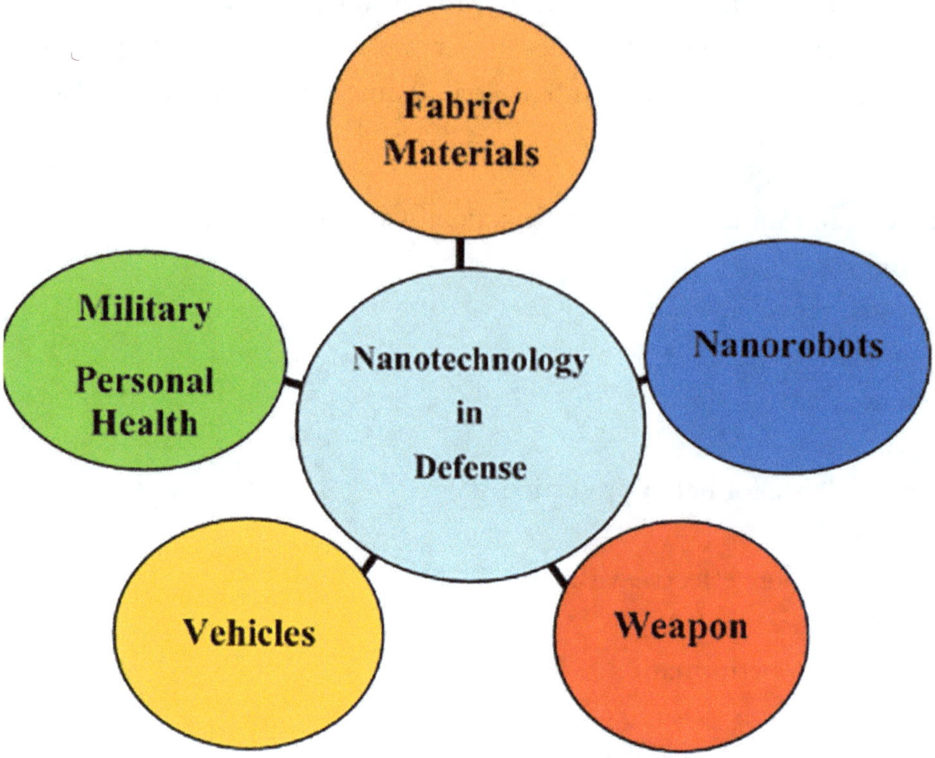

Figure 11.5 Some applications of nanotechnology in defense [12].

- *Military Clothing:* Military clothing gives protection against environment, camouflage, specific battlefield threats, flame, heat and flash, insects, chemical and the ergonomic considerations to maintain physical comfort properties. These qualities are possible only by integrating nanotechnology. Clothing worn in combat situations must be durable, crack resistant, sometimes waterproof, and must be able to operate in extreme weather conditions. For this reason, improved body armor is a major focus for military nanotechnology research. Nanomaterials like carbon nanotubes and diamond fibers have higher strength-to-weight ratios than steel, which allows for lighter and stronger armor. The Institute for Soldier Nanotechnologies (ISN) has provided an opportunity to focus funding and research activities purely on developing armor to increase soldier survival. Additional characteristics include energy-absorbing material protecting from blasts or ammunition shocks, engineered sensors to detect chemicals and toxins, as well as built in nano devices to identify personal medical issues. Fighting in extremes of hot and cold can severely restrict combat capability. This problem could be alleviated with nanowires and hydrogels which can keep soldiers cool in hot deserts and jungles as well as warm in frigid climates. Soldiers and vehicles will soon be equipped with micro antennas which can be inlaid into a soldier's uniform with nanomaterials inter-woven into the textile. Figure 11.6 show typical military uniforms [12].

Figure 11.6 Typical military uniforms [12].

- *Military Medicine:* It is important for a nation to ensure that the fighting force has the best possible medical equipment and techniques before, during, and after combat. In the field of military medicine, nanotechnology opens up new possibilities for treating injuries and diseases. Nanoparticles can be used to target drug delivery, speed up wound healing, or even create "smart" bandages that monitor the condition of a wound and release antibiotics as needed. Military medicine has been evolving over the course of millennia. The more advanced a military's wartime medical capabilities, the more likely the combatants are to survive, recover faster and more completely, and return to duty sooner. Medical improvements during the critical initial period of injury increase the chances of survival and minimize the long-term impact of injury. Improved medical technologies during rehabilitation can shorten recovery time and improve the quality of life for service members who either return to active duty or return to their civilian lives. One technology proving multiple applications to medical care during and after combat is nanotechnology. Several advantages to having access to materials on nano scale include a drastic increase in surface area to volume ratio, the ability to directly target specific tissues, and the ability to create new and novel shapes of particles. These advantages are critical in nanotechnology's medical possibilities. One of the most anticipated uses of nanomedicine is for the development of new classes of pharmaceuticals and drug delivery systems, especially for targeted therapies. Figure 11.7 illustrated the application of nanotechnology in military medicine [13].

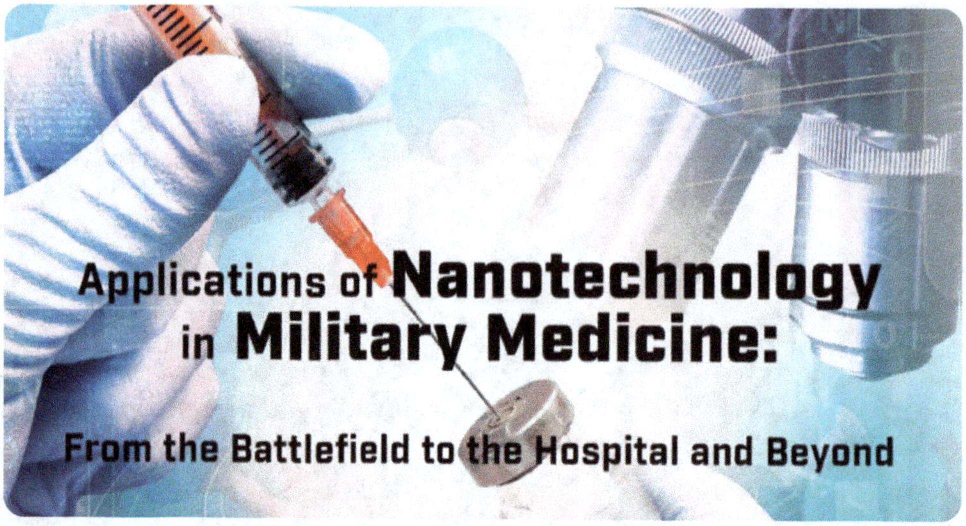

Figure 11.7 Application of nanotechnology in military medicine [13].

- *Military Weapons:* Research into military nanotechnological weapons includes production of defensive military apparatus, with objectives of enhancing existing designs of lightweight, flexible, and durable materials. These innovative designs are equipped with features to also enhance offensive strategy through sensing devices and manipulation of electromechanical properties.

- *Communications:* Traditional communication systems, such as wired and wireless systems, have limitations when it comes to military and defense applications. They are not suitable for remote locations, and they are vulnerable to interception. VSAT (Very Small Aperture Terminal) technology is the best option for military and defense applications. It offers secure, reliable, and high-speed connectivity to remote locations. Nanotechnology designed for advanced communication is expected to equip soldiers and vehicles with micro antenna rays. It facilitates easier defense related communications due to lower energy consumption, light weight, efficiency of power, as well as smaller and cheaper to manufacture.

- *Power Generation:* The electrical power sources can be integrated into the textile which remained washable. For example, nano-enabled photovoltaic, thermoelectric, and piezoelectric devices are on the cusp of being able to harvest electricity from solar energy, waste body heat, and kinetic motion from a soldier's movement. In most cases, these power sources could be embedded directly inside the uniform fabric. The common option for powering portable electronics for both military and non-military use are lithium-ion batteries due to their high energy density, long lifespan, and because they are rechargeable. Battery designers are looking at how to use graphene. The 21st century soldier is weighed down with batteries for electric laptop computers, handheld computers, night-vision goggles, and optoelectronic rifle sights. That load could soon be reduced using nanoparticles. Nanotechnology offers many ways to move electricity to the battlefield.

11.5 MILITARY NANOTECHNOLOGY AROUND THE WORLD

Nanotechnology, a field that deals with the manipulation of matter at the atomic and molecular level, is revolutionizing the military around the world, as typically shown in Figure 11.8 [14]. Industries and governments across the world are investing heavily in nanotechnology. Over the past two decades, numerous countries have funded military applications of nanotechnology including United States, China, United Kingdom, and Russia. Between 2001 and 2004, approximately 60 countries globally implemented national nanotechnology programs. The North Atlantic Treaty Organization (NATO) is intensively exploring the potential of these miniature technologies to strengthen the defense capabilities of its member states. Nanotechnology in the context of NATO represents not only a technological revolution but also a strategic imperative. We consider the application of military nanotechnology in the following nations [15,16]:

Figure 11.8 Nanotechnology revolutionizes the world of military [14].

- *United States*: The US government has been considered a national leader of research and development in the area of nanotechnology. However, US is now rivalled by international competition as appreciation of nanotechnology's eminence increases. In 2000, the United States government developed a National Nanotechnology Initiative (NNI) to focus funding towards the development of nano-science and nanotechnology, with a heavy focus on utilizing the potential of nano-weapons and coordinating application of nanotechnology among all military factions including Air Force, Army and Navy. The NNI is aimed at coordinating Federal nanotechnology research and development in the United States. Since then, the overall DoD investments (hundreds of millions of dollars) in nanotechnology each year has remained high. DoD plans to continue with this priority throughout the 21st century. The United States is the only country that has chosen to dedicate specific investments in nanotechnology-based programs for defense purposes. For the US, maintaining technological superiority constitutes a strategic advantage. DOD lists nanotech as one of 'its major strategic research programs, and spends more money on nanotechnology research than many other areas. Some examples of programs sponsored by the DOD include MIT's Institute for Soldier Nanotechnologies and the Defense University Research Initiative on NanoTechnology, which aims to support and promote nanotechnology research relevant to national defense at US universities.

- *China*: China, like the US, is focusing much of its R&D investment on military applications. In nanomaterials, China secures second place behind the United States in the amount of research publications they have released. China now produces more research papers on nanotechnology than any other nation. Conjecture stands over the purpose of China's quick development to rival the US, with 1/5 of their government budget spent on research (US$337million). In 2018, The scientific nanotechnology team at Tsinghua University, Beijing hinted at aerospace, and armor boosting applications, showing promise for defense related nano-weapons. The Chinese Academy of Science has stated the need to focus on closing the gap between "basic research and application," in order for China to advance its global competitiveness in nanotechnology.

- *Europe:* In its communication entitled "Toward a European Strategy for Nanotechnology," the European Commission estimates that the future market for products issued from nanotechnology could rise to hundreds of billions of Euros by 2010 and to one trillion thereafter. It is only to a lesser extent that ethical, societal, and health concerns about nanotechnology have been discussed. Although simple forms of nanotechnology are used in a few consumer products—like some new semiconductors, sunscreens, and stain-resistant trousers—it is not clear that such products are worth billions of taxpayer dollars. Nanotechnology will inevitably lead to disruptive technologies. If such technologies could lead to the development of a new generation of weapon systems and combatants, they could also give rise to the growth of disturbing factors affecting the global military balance. It took times before the EU realizes the necessity to develop a genuine strategy in the field.

- *Russia:* Emerging technologies are often perceived as carrying the potential to revolutionize governmental structures, economies, militaries, and entire societies. Russian leadership shares that belief. Russia has invested in nanotechnology for military purposes in a number of ways. In April 2007, then Russian president Vladimir Putin extolled nanotechnology research as the key to establishing Russia's competitive advantage in the hightech world economy and the next round of the arms race. The Russian parliament approved a $7 billion investment in nanotechnology over five years. In 2018, Vladimir Putin's decree established ERA Technopolis, a military research and development (R&D) center that focuses on nanotechnology, AI, robotics, and other technologies.

- *India:* Considering the underlying salience of nanotechnology, India has been putting in a consistent effort in the field. The potential of nanotechnology in India was realized by 2001 when the government of India set up NSTI (Nanoscience and Technology Initiative). Since then,

India has come a long way. It is carrying out extensive work in nanotechnology to enhance its application in the defense sector. DRDO has also set up a nano research and production facility in various parts of India. A Bengaluru-based Log-9 Materials startup is also collaborating with the defense industry to help it build multiple products and applications while conserving energy. However, the progress made by the country is not enough, and the process needs to be accelerated.

- *Israel:* This is the first country to publicly state they are planning to use nanotechnology in weapons. It seems to be stretching the point somewhat to suggest that no one has previously thought of sending nanotechnology to war. Several Israeli Universities such as Haifa's Technion, Jerusalem's Hebrew University and Tel Aviv University are exploring nanotechnology. Israel Aerospace Industries-Elta is teaming up with MassChallenge to develop nanotechnology set to transform both the civilian and military worlds.

11.6 BENEFITS

Nanotechnology has many benefits in the military. It offers numerous advantages, such as lighter and stronger materials, improved sensors, and targeted drug delivery systems. It can help wounds heal faster and reduce the risk of infection. For example, nanomaterials like silver and chitosan can be used to treat wounds. Artificial red blood cells and platelets can help reduce blood loss in traumatic injuries. Other benefits of military nanotechnology include the following [17]:

- *Automation:* As nanotechnology allows the further development of the "battlefield network," there will be a tendency to delegate more and more decisions to semi-autonomous systems which respond automatically to developing situations.

- *Lighter and Faster Vehicles*: Nanomaterial composites can provide added strength without extra weight. This can boost protection, increase speeds, and lower fuel consumption of aircraft, tanks, and ships. Stealth ships and aircraft are being improved with the use of nanomaterials which can help "hide" military hardware.

- *Improved Armor:* Current equipment for bulletproof and blast-proof wearable protection is reaching the limit of what a soldier can carry. Further studies have shown that nanoadditives can be used in the polymers that make up modern dressings to provide antibacterial benefits. Armor has to prevent the energy from those projectiles from being transferred to the wearer. Body armor may stop a bullet from piercing the skin, but the energy transferred to the body from the impact may still kill the wearer.

- *Medical Care:* Nanotechnology is an emerging technology that has the potential to revolutionize military medical care. Nanotechnology can help deliver drugs and other medicines quickly and accurately. It can help monitor a soldier's health, such as their heart rate and blood pressure, using non-invasive technologies. It can also help with advanced medical monitoring and diagnostics.

- *Wound Treatment:* The threat of injury and even death hangs over the head of most active men and women in the armed forces. The treatment for some injuries can be life-threatening as well. Wound treatment in long-term care settings can be quite different than treatment on the battlefield. Nanomaterials of copper and silver have been proven to have antimicrobial effects which when applied to bandages can help to keep wounds free of infection. Nanotechnology also offers the promise of creating more efficacious vaccines and anti-infectives, which could be deployed easily to combat areas and remote locations. Figure 11.9 shows a wounded soldier [12].

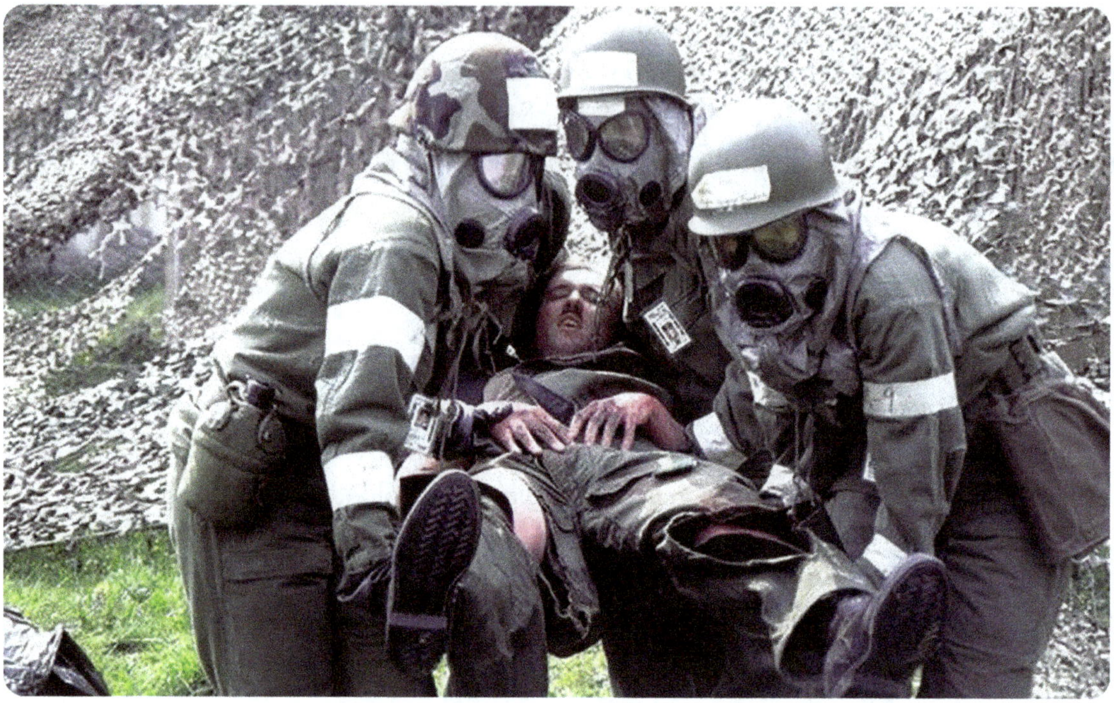

Figure 11.9 A wounded soldier [12].

- *Next Generation Drones:* Nanotechnology is finding ways to make ever smaller electronics which could be used to build miniature drones. It is possible that a large-scale fleet of tiny (and therefore hard to shoot down) drones could provide a destructive force if enough drones were applied to a specific high value target.

- *Soldier Protection*: Nanotechnology for the soldier is directly related to new functionalities in his suit, helmet or other portable equipment. The future war-fighter is equipped with powerful tools to cause drastic damage or to neutralize the opponents. He is in a smart uniform and smart helmet, which has sensors to protect him from ballistic elements. He identifies objects or enemies through RFID tags and a body area networks, consisting of a number of wireless products communicating with each other. The war-fighter has the ability to monitor his position, his physical and mental condition, and status of equipment.

11.7 CHALLENGES

While there are many benefits to military nanotechnology research, there are also several challenges which should be handled with great care as the technology progresses. It is vital not to overlook the role of the military in the development and dissemination of potentially harmful nanotechnology applications. Other challenges of military nanotechnology include the following [14,15]:

- *Regulation:* International regulations are non-existent for issues of nanotechnology and its military applications. There will be an enormous challenge to regulate the use of nanobots, miniature communications systems, etc. Some of the medical applications of nanotechnology, developed to improve solders' endurance and performance, would also need careful regulation for general medical applications. Ambiguity and a lack of transparency in research increases difficulty of regulation in this area. Producing legislation to keep-up with the rapid development of products and new materials in the scientific spheres, would pose as a hindrance to constructing working and relevant regulation.

- *Security:* Nanotechnology poses grave risks for international security and future military balances. Widespread availability of the nanotechnology-based devices would inevitably lead to their use for criminal activity and terrorist attacks. Although nanotechnology and its use in warfare promise economic growth, the promise comes with the increased threat to international security and peacekeeping. Such developments will have impact on geo-politics, ethics, and the environment.

- *Safety:* It is important that consumers and workers are ensured proper protection and safety when nanotechnology is used in the military. The introduction of nanotechnology into everyday carries the possibility of unknown consequences for the environment and safety. Convergences between nanotechnologies and biotechnologies should lead to new therapeutic agents of greater specificity and safety.

- *Environment:* While there are several positive environmental effects of nanotechnology, there are certain negative impacts as well which need to be addressed, such as increased toxicological pollution of the environment due to the uncertain shape, size, and chemical compositions of some of the nanotechnology products or nanomaterials. The release of toxic nano-substances can cause environmental harm. Associated risks may involve military personnel inhaling nanoparticles added to fuel, possible absorption of nanoparticles from sensors into the skin, water, air or soil, dispersion of particles from blasts through the environment, alongside disposal of nano-tech batteries potentially affecting ecosystems. There can be some risks involved with engineered nanomaterials, which can also negatively impact the environment. For example, military activities can result in explosions; blasts by high-tech weaponry can release toxic nanoparticles.

- *Socio-ethical Concerns:* When discussing military applications of nanotechnology, it is equally important to discuss the social, environmental, and ethical concerns as well. It is unknown the full extent of consequences that may arise in social and ethical areas. The main ethical uncertainties entail the degree to which modern nanotechnology will threaten privacy, global equity, and fairness. An overarching social and humanitarian issue, branches from the creative intention of these developments. "The power to kill" highlights the unethical purpose and function of destruction these nanotechnological weapons supply to the user. The application of nanotechnology in certain areas of defense may in the future be deemed unethical or may simply upset the balance of power by creating a super-weapon or by rendering an opponent's weapons impotent.

- *Human Rights Concerns:* These concerns related to nanotechnology include uneven access to the benefits of technology, unfair distribution of the profits associated with technology, negative social and anthropological impact of the technology, privacy and informed consent.

- *Secrecy:* The secrecy behind military research and nanotechnology will make enforcing legislation impossible. It is impossible to determine if a nation is non-compliant if one is unable to determine the entire scope of research, development, or manufacturing. It is therefore difficult to know exactly where the technology is heading or what powers the weapons of the future will possess.

- *Autonomous Systems*: The integration of artificial intelligence and autonomous systems (such as drones and unmanned ground vehicles) into military defense has the potential to revolutionize warfare. Global tensions continue to rise as several nations invest heavily is the integration of autonomous weapons and nanotechnology into their defense strategies. Autonomous weapons are becoming increasingly prevalent in military operations around the world. For example. US and China have been investing heavily in autonomous weapons. The use of nanotechnologies in autonomous weapon systems should be strictly regulated by international agreements. It is crucial to ensure that the decision to use force always remains under direct human control, with clear accountability protocols. Figure 11.10 shows some autonomous systems [18].

Figure 11.10 Some autonomous systems [18].

- *Surveillance:* A robust legal and ethical framework is needed to balance the potential of nanotechnology for surveillance and intelligence gathering with the right to privacy. This framework should include clear rules for the use of nanotechnologies in security and intelligence, with an emphasis on protecting civil rights.

- *Asymmetric Warfare:* Military and security forces should develop comprehensive strategies and countermeasures for the potential misuse of nanotechnology in asymmetric conflicts. This includes the development of detection systems, protective measures, and tactics to counter nanotechnology-based threats.

11.8 CONCLUSION

Technology has always played a crucial role in military defense, and its importance has only increased in today's modern world. Nanotechnology is the understanding and control of matter at the nanoscale (roughly 1-100 nm), where unique phenomena enable novel applications. It is generally accepted that advances in nanotechnology will drive the next paradigm shift in science and technology. As with most new, emerging technologies, military applications of nanotechnology are likely to be the first to be realized. It appears likely that most military technology will be dependent on nanomaterials.

Currently, nanotechnology is evolving from the basic stage of its development into the applied research stage of technology maturity. It is highly promising prerequisite for military applications. It has not yet acquired the status of a sufficiently mature technology to allow analysts to foresee what could be its precise impact on political-military affairs in the coming decades. More information about nanotechnology and nanomaterials in the defense industry can be found in the books in [19-28] and the following related journals/magazines:

- *Nanotechnology*
- *Nanoscale.*
- *Nano: The Magazine for Small Science*
- *Micro and Nano Technologies*
- *Nanotechnology News*
- *Nature Nanotechnology*
- *Current Research in Nanotechnology*
- *American Journal of Nanotechnology & Nanomedicine*
- *Nanomedicine: Nanotechnology, Biology and Medicine*
- *Journal of Nanotechnology*
- *Journal of Nanoparticle Research*
- *Journal of Bioelectronics and Nanotechnology*
- *Journal of Nanoscience and Nanotechnology,*
- *Journal of Micro and Nano-Manufacturing*
- *Journal of Nanoengineering and Nanomanufacturing*
- *Nanotechnology and Precision Engineering*

REFERENCES

[1] P. A. Adekunte, M. N. O. Sadiku, and J. O. Sadiku, "Nanotechnology in the Military," *International Journal of Trend in Scientific Research and Development,* vol. 8, no. 5, September-October 2024, pp. 953-964.

[2] M. N. O. Sadiku, M. Tembely, and S.M. Musa, "Nanotechnology: An introduction," *International Journal of Software and Hardware Research in Engineering,* vol. 4, no. 5, May. 2016, pp. 40-44.

[3] "Nanotechnology white paper,"
https://www.epa.gov/sites/default/files/201501/documents/nanotechnology whitepaper.pdf

[4] M. N. O. Sadiku, Y. P. Akhare, A. Ajayi-Majebi, and S. M. Musa, "Nanomaterials: A primer," *International Journal of Advances in Scientific Research and Engineering,* vol. 7, no. 3, March 2020, pp. 1-6.

[5] U. A. Ali et al., "A state-of-the-art review of the application of nanotechnology in the oil and gas industry with a focus on drilling engineering," *Journal of Petroleum Science and Engineering,* vol. 191, August 2020.

[6] K.R. Saravana and R. Vijayalakshmi, "Nanotechnology in dentistry," *Indian Journal of Dental Research*, November 2005.

[7] N. S. Giges, "Top 5 trends in nanotechnology," March 2013,
https://www.asme.org/topics-resources/content/top-5-trends-in-nanotechnology

[8] D. E. Effiong et al., "Nanotechnology in cosmetics: Basics, current trends and safety concerns—A review," *Advances in Nanoparticles*, vol. 9, 2020, pp. 1-22.

[9] S. Sengupta, "Nanotechnology and the military," November 2021,
https://sustainable-nano.com/2021/11/11/nanotechnology-and-the-military/

[10] W. Soutter, "Nanotechnology in the military," June 2012,
https://www.azonano.com/article.aspx?ArticleID=3028

[11] M. S. Abed and Z. A. Jawad, " Nanotechnology for defence applications," in N. M. Mubarak, S. Gopi, and P. Balakrishnan (eds), *Nanotechnology for Electronic Applications. Materials Horizons: From Nature to Nanomaterials.* Singapore: Springer, 2022.

[12] S. Hilton, "Nanotechnology and tomorrow's infantryman," April 2023,
https://blog.polymernanocentrum.cz/nanotechnology-and-tomorrows-infantryman/

[13] "Applications of nanotechnology in military medicine: From the battlefield to the hospital and beyond," https://hdiac.org/articles/applications-of-nanotechnology-in-military-medicine-from-the-battlefield-to-the-hospital-and-beyond/#:~:text=Each%20category%20has%20direct% 20applications,hemorrhage% 20control%2C%20wound%20management%2C%20tissue

[14] "Another potential aspect of the military revolution: Nanotechnology as a key element of NATO's future," September 2024, https://www.czdefence.com/article/another-potential-aspect-of-the-military-revolution-nanotechnology-as-a-key-element-of-natos-future#:~:text=Nanotechnology%20allows%20for% 20the%20development,of%20 complex%20early%20warning%20systems.

[15] "Nanotechnology in warfare," *Wikipedia*, the free encyclopedia, https://en.wikipedia.org/wiki/Nanotechnology in warfare

[16] A. De Neve, "Military uses of nanotechnology and converging technologies: Trends and future impacts," http://www.irsd.be/website/images/livres/focuspaper/FP08.pdf

[17] "Nanotechnology and the military: How tiny materials can win wars," https://blog.nanochemigroup.cz/nanotechnology-and-the-military-how-tiny-materials-can-win-wars/

[18] S. Alsabi, "'The rise of autonomous weapons and nanotechnology in military defense: A global perspective' by AI," February 2023, https://www.linkedin.com/pulse/rise-autonomous-weapons-nanotechnology-military-defense-sultan-alsabi

[19] S. Tomar, *Nanotechnology: The Emerging Field for Future Military Applications.* Institute for Defence Studies & Analyses, 2015.

[20] D. Ratner and M. A. Ratner, *Nanotechnology and Homeland Security: New Weapons for New Wars.* Prentice Hall/PTR, 2004.

[21] M. N. O. Sadiku, S. M. Musa, T. J. Ashaolu, and J. O. Sadiku, *Applications of Nanotechnology.* Gotham Books, 2023.

[22] J. Altmann, *Military Nanotechnology: Potential Applications and Preventive Arms Control (Contemporary Security Studies).* Routledge, 2005.

[23] M. A. Medina et al., *Nanotechnology in the Space and Military Industries: A review of trends. (Nanotechnology trends).* Independently Published, 2020.

[24] R. K. Sharma, *Military uses of nanotechnology.* Sumit Enterprises, 2014.

[25] N. Kumar and A. Dixit, *Nanotechnology for Defence Applications.* Springer, 2019.

[26] M. Sharon et al., *Nanotechnology in the Defense Industry: Advances, Innovation, and Practical Applications (Advances in Nanotechnollogy & Applications).* Wiley-Scrivener, 2019.

[27] L. A. Del Monte, *Nanoweapons: A Growing Threat to Humanity.* Potomac Books, 2017.

[28] M. Kosal, *Nanotechnology for Chemical and Biological Defense.* New York: Springer, 2009.

CHAPTER 12

GAMIFICATION IN THE MILITARY

"Gamification is the use of game elements and game thinking in non-game environments to increase engagement and improve better targeting."
– Mehreen Siddiqua

12.1 INTRODUCTION

Game has become a significant part of human's culture. Today, gaming has become far more mainstream than ever before. Games have incredible power. They have proven that they can challenge society's preconceptions, change lives, and even revolutionize our society. Games tend to create intense motivation and authentic engagement for participants. They have evolved in graphic integrity, sophistication, and technological design [1]. Most games are now accessible on all sorts of devices. The US military uses video games such as America's Army and Full Spectrum Warrior as training tools. The reasons for this use include cost-effectiveness, safety and security, customization and flexibility, realism, efficiency, repeatability, and accessibility. The military continues to experiment the use of video games for training purposes [2].

Gamification is applying game mechanics and game design techniques to engage and motivate people to achieve their goals. It taps into the basic human desires and needs of the users' impulse which revolves around the concept of status, achievement, competition, and reward. Game elements are the building blocks of gamification. Although gamification does not equal "playing games," everyone should play games a few times a month. Personal digital devices has made educational games accessible to everyone. The uses of gamification are universal and can be applied in any situation. Gamification has become a new trend in the game industry [3]. This phenomenon is taking the world by storm and the American Armed Forces are no exception. Figure 12.1 shows typical video games [4].

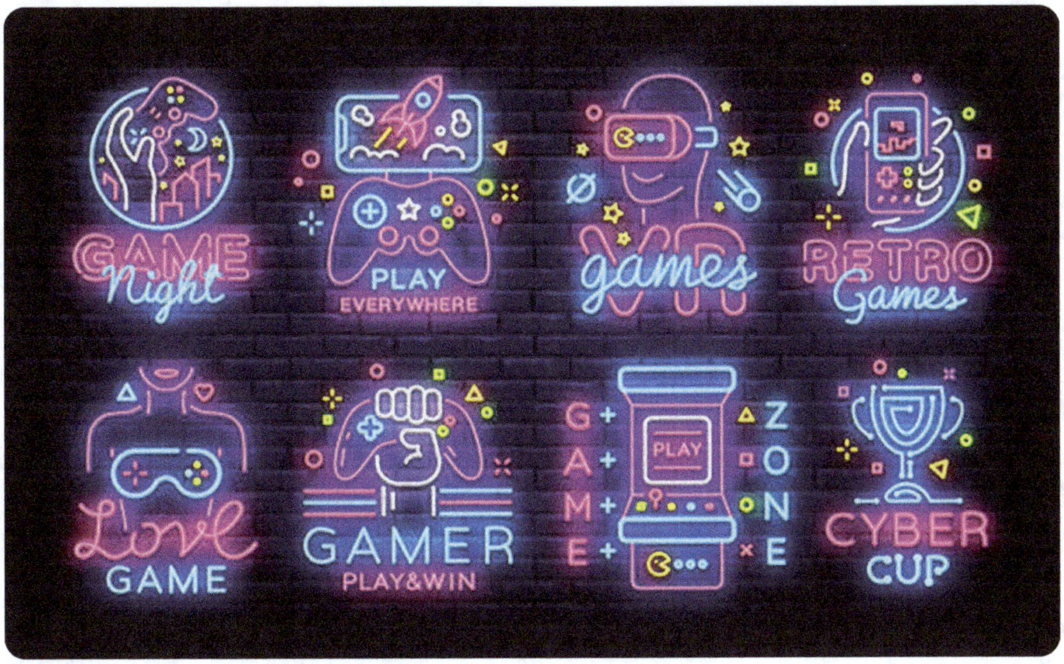

Figure 12.1 Video gaming [4].

The gaming companies have invested millions to develop user-friendly videos games. To succeed, the video game industry must offer highly advanced technologies at affordable prices. This is partly why they are so attractive to the military. The military has long been an object of the video game industry's fascination [5].

In gamification, elements of game design are used to make an experience more engaging in a non-game context such business and the military. The role of gamification in military training is transforming the way our armed forces prepare for the modern battlefield. Gamification in the military training programs is becoming more engaging, effective, and efficient. We can use gamification to boost training and take individual differences into account [6].

This chapter provides an overview on gamification in the military. It begins by describing what gamification is all about. It addresses gamification in the military and provides some applications. It highlights the benefits and challenges of gamification military. It covers global gamification military The last section concludes with comments.

12.2 WHAT IS GAMIFICATION?

The word "gamification" was coined in 2002 by Nick Pelling, a British inventor, but it did not gain popularity until 2012. The idea of gamification came from the fact that the gaming industry was the first to master human-focused design and we are now learning from games. Gamification is not a new concept, but it is deeply rooted in marketing endeavors, such as points cards, grades, and degrees, and workplace productivity [7]. Researchers became interested in gamification because the concept could be implemented in different ways to motivate people. Gamification has become hugely popular in all walks of life, including education, business, and workplace.

The gamification can be viewed in two ways: (1) adopting the act of playing a video game into everyday use, (2) the act of using game elements to make non-games more enjoyable. It is applied in education, business, sports, marketing, and finance. It is currently one of the largest trends in education. Traditional education has been found to be ineffective in motivating and engaging many students. Gamification is cutting-edge approach which is producing positive results in every region of the world.

Everyone loves games. Gamification just takes advantage of that innate desire. It can make practice fun. It can make the routine less droll. Online games have become bolder and more diverse. Gamification involves the introduction of gameplay to a traditionally non-game environment. Its goal is to encourage user engagement [8].

The most popular gamification features include [9,10]:

- *Points:* They represent progress and accomplishment; they are often used to obtain rewards. Learners earn points for completing tasks or achieving a certain result.
- *Leaderboards*: Learners can compete to earn the most points and rank highest out of their peers. Leaderboards can be an exciting and dynamic addition to the learning experience.
- *Badges:* They are similar representation to points, often also representing a status. Reward completion, good results, and exemplary behavior with virtual badges that learners can accumulate on their profile.
- *Rewards:* Beyond badges and points, real-life rewards like vouchers and discounts can further incentivize participation.

These elements can be used to gamify your app.

12.3 GAMIFICATION MILITARY

War is a nightmare. It is a violent means of achieving a political objective, which is mostly self-centered. The defense industry is asking for reduction in the number of humans needed in combat. For example, DARPA shared that they want more artificial intelligence, consisting of ground robots operating alongside warfighters. The robot can do this all without suffering or death. They are taking soldiers out of harm's way and increasingly replacing them with more unmanned combat vehicles [11]

Military institutions worldwide have been bringing innovations for many years. As illustrated in Figure 12.2, the military career is surrounded by gamification all the way [12]. Few workplaces can boast a culture as the military. Solders go through ranks: Captain, Colonel, Major, Commander, Brigadier General.

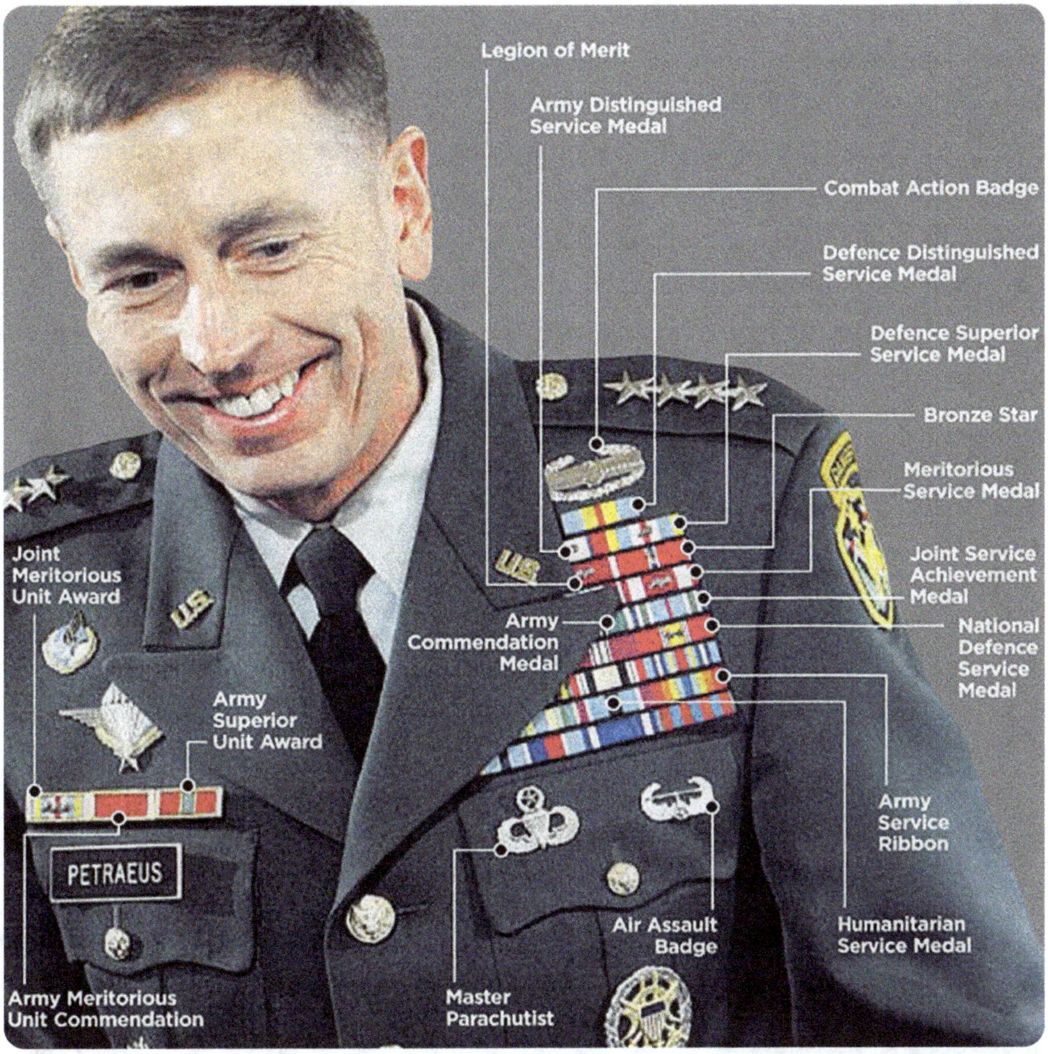

Figure 12.2 The military career is surrounded by gamification all the way [12].

Gamification and 3D training significantly increase combat effectiveness by increasing the level of training provided to military personnel who operate the equipment. They are being used by the Army to address training challenges including providing immersive training for the tanks. The best part of gamification in the military is how much of it can now be done remotely. In 2008, the US Army invested $50 million over five years on games and gaming systems designed to prepare soldiers for combat.

12.4 APPLICATIONS OF MILITARY GAMIFICATION

Gamification involves taking some of the things that make gaming attractive, addictive, interesting and challenging, and applying them to those non-gaming areas such as business, education, etc. The military has been utilizing gamification solutions for its recruits to digest and assimilate information more quickly and efficiently. A lot of this upskilling in the military comes down to physical training. The following are just typical applications of gamification in the military.

- *Training*: When it comes to military training, gamification is a great match made in heaven. The use of serious games in military training has become a force to be reckoned with. Video games can be used to gamify all areas of training, such as language and cultural training, making it more engaging and effective. Serious games offer a risk-free space for trainees to experiment, make mistakes, and learn without real-world consequences. They can be used to teach a wide range of skills, from language and cultural understanding to logistics and strategy. Gamification can boost soldier training, while taking individual differences into account. Figure 12.3 displays some soldiers in training [13]. The following factors make gamification such a powerful tool for military training [14]:

Figure 12.3 Some soldiers in training [13].

  - *Enhanced motivation:* Gamification taps into our innate love for competition, making learning more engaging and enjoyable.

  - *Accelerated learning:* By breaking down complex concepts into digestible, game-like experiences, trainees can grasp new skills more quickly.

  - *Immediate feedback:* Gamified environments offer real-time feedback, allowing trainees to learn from their mistakes and make adjustments on the fly.

- *Recruitment:* Gamification in recruiting is buzzing in the human resource department. We could gamify certain elements of the recruitment process. Gamification is not new to recruitment. For example, the French Post Office had problems retaining people. So it gamified what it is like to be a post office worker. The result was that it allowed people to self-filter; those that actually got in did stay. The military aimed to replicate success of the French Post Office. Gamification Nation designed a gamified application to simulate the necessary skills and tasks to make the person more informed about the demands of the role. This ultimately motivated the right people to go in to the Royal Navy [15]. The America's Army video game has long been used as a recruitment and training tool. The US military came up with an ingenious idea to recruit the most talented prospects. In hiring, gamified assessments are used to discern a candidate's personality traits. With customized recruitment gamification tools, employers can attract a broad range of candidates, as shown in Figure 12. 4 [16]. Using gamification in human resources is a win-win solution for both employers and the candidates.

Figure 12.4 A broad range of candidates that employers can attract using gamification [16].

- *Learning and Development:* The military has always been an advocate of using games in learning. So they use all kinds of game-based techniques for learning, and development, thereby upskilling in the military. The military has been utilizing gamification solution for its recruits to digest and assimilate information more quickly and efficiently. Gamification mixed with remote learning is extremely helpful for medics and military engineers. Gamified learning ensures realism in virtual training and ensures more knowledge sharing. It can be applied to any training needed to include learning how to operate a vehicle. Instead of instructors teaching students all day, they are moving from passive to active learning experiences in the classroom. Figure 12.5 show the learning gamification options [17].

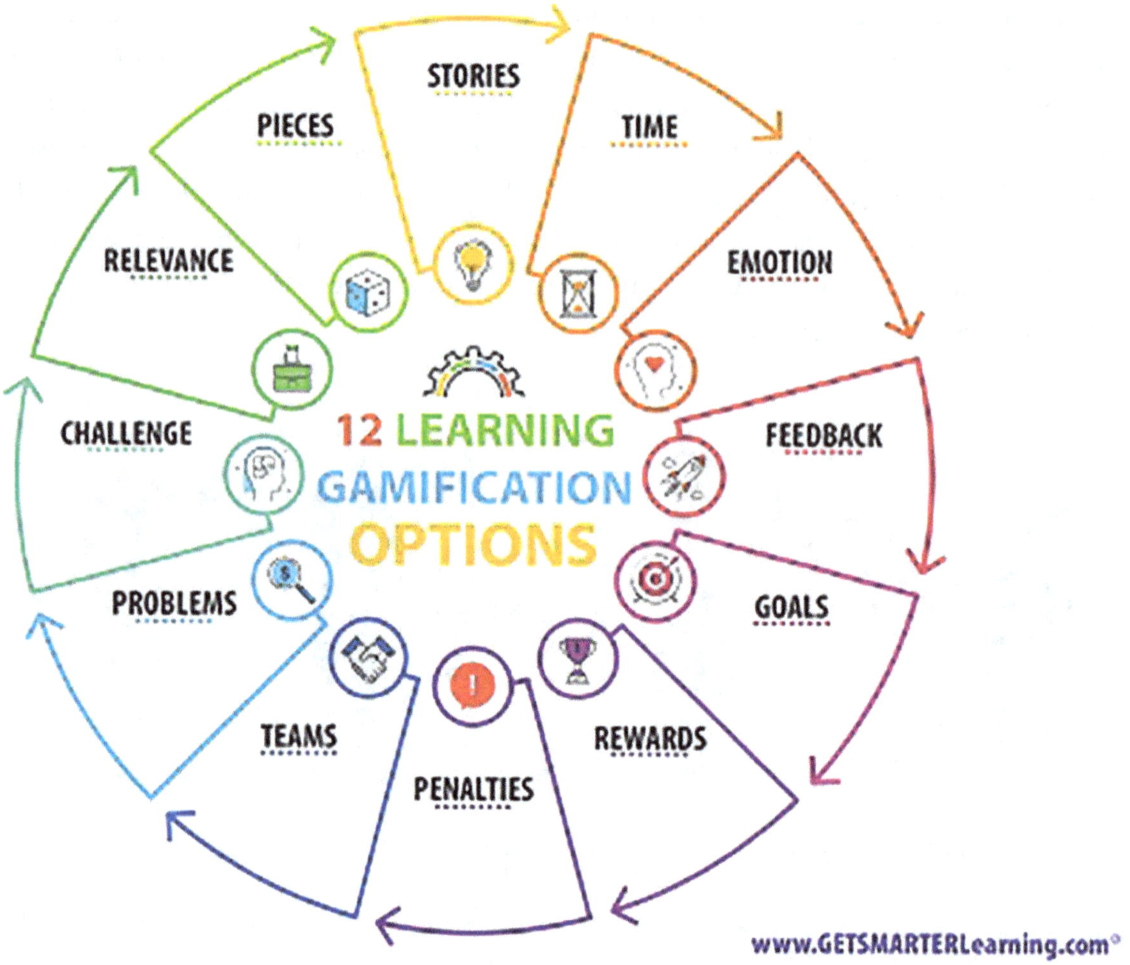

Figure 12.5 The learning gamification options [17].

- *Simulations:* The use of simulations could be regarded as the "original" and oldest form of gamification in the military. Simulation is an approach the military uses to test people's ability to handle high-end equipment like fighter planes. Simulation game training is heavily integrated because it allows for soldiers to experience a virtual reality with realistic combat situations without any of the dangers. The same simulation technology has been used to remove soldiers from danger entirely. In the past, the military focused mostly on more physical simulations. However, modern-day simulations can also be accessed via mobile devices. The US military is using a wide variety of video games and simulations, including virtual reality, augmented reality, and other advanced technologies, to train soldiers. Military officers can use remote gamification to practice simulations where they handle managing resources.

- *Combat Readiness:* For the military, readiness for combat is the number one priority. Defense Agencies worldwide rely on well-trained teams to maintain their strategic advantage on the battlefield through combat readiness. The Ukraine war has shown how critical training is to combat readiness. The gamification of training is a new tool used to support readiness for combat [18]. Figure 12.6 shows some soldiers in combat [12].

Figure 12.6 Some soldiers in combat [12].

12.5 BENEFITS

When it comes to the benefits of gamification, the sky is the limit. Gamification can incorporate many forms of recognition like military awards such as increased pay, ribbons, awards, badges of honor, and medals. It is the best way to hammer in the essential military skills. It increases motivation when workers are given rewards and receive validation for their efforts. Gamified learning makes training realistic, accurate, dynamic, visual, and available in real-time, which is critical for today's mission readiness goals. We can use gamification to boost training, while taking individual differences into account. Also, gamification in human resources can project an innovative image of an employer, reveal people's interest in job openings, and predict future job performance of the applicants.

Other benefits of gamification in the military include [19]:

- Gamification provides an engaging learner experience, leading to behavior change that motivates learners
- It offers an immersive learning experience which does wonders for motivating people to aim for peak performance and positive behavioral change
- It has been a driving force to help enlist recruits
- It is widely used for learning and development in the armed forces
- It can give people the validation they need to maintain high-level sharpness and success
- It is being used to attract, select, on-board, and develop employees
- It keeps up the team spirit
- 24/7 access to individual training for personnel to practice in a realistic, virtual environment
- Immediate remote support by experts around the globe who are also online and using gamification and 3D training software

12.6 CHALLENGES

Video games have been blamed for many reasons in our society. They have contributed to violence in American society. Games can get their hooks into people. A criticism of massive-multiplayer games is that it encourages players to remain inside playing rather than outside exercising. The same applies to the simulations the military has relied on.

Gamification does not reliably lead to better training outcomes. Some organizations have not embraced gamification due to a lack of clarity about what it can do for them, how to implement it, and how it benefits them. A lot of HR professionals and their organizations are skeptical about embracing gamification recruitment tools.

Gamified instruction and education is not ideal for all students. Groups that received gamified training showed mixed results in short-term material retention compared with the control groups. Since mandatory play may limit the benefits of gamified learning, voluntary participation is a key tenet for success. Researchers in the defense enterprise must make sure that the potential of game-based learning is a reality for military education [20].

12.7 GLOBAL GAMIFICATION IN MILITARY

Military institutions worldwide have been bringing innovations for years. When it comes to military training, gamification is a match made in heaven. The potential target audience for gamified applications is large. Gamification has been a driving force to help the military enlist recruits. It is also widely used for learning and development in the armed forces. As a promising learning method, gamification is a growing trend in manufacturing, education, healthcare, military training, etc. The role of gamification in military training is transforming the way our armed forces prepare for the challenges of the modern battlefield. As technology evolves, so does the way we prepare our armed forces for future challenges [21]. We consider how the United States, Europe, and China incorporate gamification in their military.

- *United States*: Without training drills and skill tests, the military would not be capable of much. Scientists and trainers have used gamification to make training more efficient and effective. Improving training supports the Army's modernization priority for soldier lethality. The role of gamification in military training has become increasingly vital in preparing our soldiers for real-life combat situations. Gamification could improve military training and learning process by showing blended, immersive, and interactive around-the-clock training [19]. In 2002, the US Army released the online role-play game design to provide civilians with an inside perspective and virtual role in today's premier land force: the U.S. Army. The game provided players with a chance to participate in a number of military occupations. The recruitment game America's Army has retired after 20 years [22]. The US Department of Defense is exploring the potential of "gamified learning" to enhance education in fields such as foreign language study.

- *Europe:* The Pizarro simulators are expected to become among Europe's largest joint armored vehicle training networks and the most advanced simulation centers internationally. Indra has begun supplying its Pizarro infantry combat vehicle simulator to multiple Spanish Army bases. These compartments project provides soldiers with a fully-immersive training experience. They allow the soldiers to quickly prepare for their missions and learn how to communicate and coordinate, where to move,and how to react in the face of the enemy. It also capitalizes on the new technologies linked to virtual reality and gamification and maximizes the communication and interoperability capabilities of the systems. The Pizarro vehicle simulators are shown in Figure 12.7 [23].

Figure 12.7 The Pizarro vehicle simulators [23].

- *China:* China has always viewed video games as a negative influence to society and the social order. However, China has designed a video game for military training purposes - Glorious Mission, as a tool for the US military and also as a lure for army recruitment. Since its public release, Glorious Mission, a video game initially designed for and by the Chinese military, has been downloaded more than a million times. It was designed as a training aid for Chinese soldiers and controlling virtual People's Liberation Army troops in various battlefield scenarios. The decision to make the game available to the wider public was taken in order to instill patriotic values in its players. The game will help also drum up new army recruits. Online games might be a powerful tool for influencing thoughts and ideas and making young boys desire to become a soldier [24]. Figure 12.8 shows Glorious Mission [25].

Figure 12.8 Glorious Mission [25].

12.8 CONCLUSION

Gamification is all around us. It is the art of incorporating game-like elements into non-game contexts such as education, business, and warfare. As technology continues to evolve, one can expect gamification to play a more prominent role in the future of military training. Gamified training will make soldiers better equipped to handle the rigors of real-world combat scenarios. Gamification will always be relevant and effective.

Gamification is here to stay and will become more widespread in the coming years. The marriage of gamification and military training is a win-win scenario. In the future, warfare will be increasingly fought by engineers, scientists, accountants, and politicians. Accountants and politicians will decide if they can afford more unmanned vehicles. For more information about military gamification, one should consult the books in [26-28] and the following related journals: *Journal of Games, Games Art, and Gamification* and *National Defense Magazine.*

REFERENCES

[1] N. King, "6 Gamification trends that will transform training in 2020 & beyond," May 2020, https://nicoking.medium.com/6-gamification-trends-that-will-transform-training-in-2020-beyond-d0d7f044a29c

[2] "Did you know video games are being used for military training?" https://medium.com/illumination/did-you-know-video-games-are-being-used-for-military-training-fd563dc7a5c5#:~:text=The%20military%20has%20been%20using%20different%20types%20of%20games%20for,critical%20thinking%2C%20and%20teamwork%20skills.

[3] "Gamification 101: The rise of game-everything," January 2018, https://medium.com/@AgateInt/gamification-101-the-rise-of-game-everything-b0b2dc324e80

[4] https://stock.adobe.com/search?k=video+game

[5] "The military is weaponizing video games," https://theweek.com/articles/450858/military-weaponizing-video-games

[6] . M. N. O. Sadiku, U. C. Chukwu, and J. O. Sadiku, "Gamification in the military," *International Journal of Human Computing Studies*, vol. 5, no. 5, May 2023, pp. 21-29.

[7] K. Seaborn and D. I. Fels, "Gamification in theory and action: A survey," *International Journal of Human-Computer Studies*, vol. 74, 2015, pp. 14–31.

[8] M. N. O. Sadiku, S.M. Musa, and R. Nelatury, "Digital games,*" International Journal of Research and Allied Sciences*, vol. 1, no. 10, Dec. 2016, pp. 1,2.

[9] V. V. Abraham, "Gamification: A digital marketing strategy in the beauty industry," October 2020, https://www.linkedin.com/pulse/gamification-digital-marketing-strategy-beauty-industry-abraham

[10] C. Pavlou, "How to use gamification in the workplace: Tips and examples," November 2020, https://www.talentlms.com/blog/tips-gamification-workplace/

[11] K. Benedict, "The future of warfare, its purpose and gamification," June 2021, https://www.linkedin.com/pulse/future-warfare-its-purpose-gamification-kevin-benedict

[12] D. Goudet, "Gamification in the military," November 2016, https://www.linkedin.com/pulse/gamification-military-david-goudet

[13] H. Jacobs, "The Pentagon has selected 2013's most intense military training photos," January 2021, https://www.businessinsider.com/photos-of-us-military-training-2014-1

[14] "The role of gamification in military training: Serious games and their applications," April 2023, https://www.warfighterpodcast.com/blog/the-role-of-gamification-in-military-training-serious-games-and-their-applications/#:~:text=Gamification%3A%20A%20Game%20Changer%20in %20Military%20Training&text=Enhanced%20motivation%3A%20Gamification%20taps%20into,grasp%20new%20skills%20more%20quickly.

[15] H. Dudfield, "Gamification to support the military," October 2020, https://www.qinetiq.com/en/blogs/gamification-to-support-the-military

[16] D. Wali, "Gamification in recruiting: Everything you need to know," https://thetalentgames.com/gamification-in-recruiting/

[17] "Advanced principles of learning gamification training," https://www.webagesolutions.com/courses/TP2980-advanced-principles-of-learning-gamification

[18] "Gamification and 3D training are important new tools that defence is using to support readiness for combat," https://soucy-defense.com/gamification-of-training/

[19] "Gamification in the military: A new way of shaping the heroes of tomorrow," https://smartico.ai/gamification-in-military/

[20] D. J. Finkenstadt and E, Helzer, "Gamified learning can be effective," March 2023, https://www.usni.org/magazines/proceedings/2023/march/gamified-learning-can-be-effective

[21] "The role of gamification in military training: Serious games and their applications," April 2023, https://www.warfighterpodcast.com/blog/the-role-of-gamification-in-military-training-serious-games-and-their-applications/

[22] K. Kapp. "US Army's online recruitment game sunsets after 20 years," https://www.linkedin.com/pulse/us-armys-online-recruitment-game-sunsets-after-20-years-karl-kapp

[23] R. Manuel, "Spain receives pizarro infantry combat vehicle simulator," May 2023, https://www.thedefensepost.com/2023/05/22/spain-pizarro-training-simulator-indra/

[24] J. Sudworth, "Why China's military has turned to gaming," April 2013, https://www.bbc.com/news/world-asia-china-21999036

[25] "Glorious Mission – DramaWikui,: https://wiki.d-addicts.com/Glorious Mission

[26] D. Noh, *The Gamification Framework of Military Flight Simulator for Effective Learning and Training Environment*. University of Central Florida, 2020.

[27] A. Chapman, *Can Gamification Save Unions? An Investigation Into the Use of Gamification by Military and Union Movements : a Dissertation Submitted to Auckland University of Technology in Partial Fulfilment of the Requirements for the Degree of Bachelor of Business with Honours*, 2015.

[28] L. Leenen and N. van der Waag-Cowling (eds.), *Proceedings of ICCWS 2019 14th International Conference on Cyber Warfare and Security*. Academic Conferences and Publishing Limited, 2019.

INDEX

www.ingramcontent.com/pod-product-compliance
Lightning Source LLC
LaVergne TN
LVHW081533070526
838199LV00005B/351